排版的风格

TYPOGRAPHIC STYLE
CONSTRUCTIONAL STYLE, INTERNATIONAL STYLE,
CLASSICAL STYLE & GRID SYSTEM

左佐 编著

电子工业出版社·
Publishing House of Electronics Industry
北京·BEIJING

目录

Preface 序言　7

1　Classical style 古典风格　12
2　Constructional style 构成风格　48
3　International style 国际风格　74
4　Popular style 流行风格　102
5　Western grid system 西文网格　110
6　Chinese grid system 中文网格　140

序言

排版的技术与风格

什么是排版（Typography）？Typeface、Lettering、Layout、Compsition？*Modern Typography* 这本书中，认为当第一本字体排印手册（Type Specimen）出现的时候，现代排版就开始了。之前排版只是印刷技术中的一个环节，由印刷工人根据功能和经验完成。但有了排印手册后，美学开始介入，比如我们都知道标题要比正文大，这是功能的需要，但是到底大多少则不仅需要技术，还需要审美。

所以排版有两个部分：基于功能的技术和基于审美的风格。功能技术部分有对错之分，比如"，"位于行首，并不会有损美观，而是有悖功能，是不符合规范的。这部分的应用有三种情况：正确使用排版的功能技术——专业；明知正确用法而故意非正确使用——实验；不知正确用法而错误使用——无知。

对于以阅读体验为导向的长文书籍，一般会很在意功能技术上的对错；这些功能技术大部分是烦琐的成文规定。所以本书意不在此，而是更偏重排版的视觉风格。所谓风格可以说是一人创制、多人仿制形成的一股趋势，这"一人"必须著名且艺高。

排版的逻辑线

我们一般如何开始排版？上设计网站搜关键词，找参考资料，借鉴套用最近流行的？翻阅年鉴作品集，看到喜欢的借鉴套用？分析图文，制作适合的网格，把图文放进去？这些看似都是可行的方法，但都缺乏逻辑。

为什么很多设计师看了大量的排版书籍或文章，还是觉得书架上缺一本排版书籍？因为大部分书籍或文章是在传授零碎的排版经验，这种情况行距大一些，那种情况标题小一些，这个标点应该缩进，那个符号后要加空格等。当碰到命题排版时，大脑调取的都是随机出现的碎片，不稳定且混乱，没有确切的原因标准，不知从何着手，于是迫使自己再去看相关的知识来填充。其实真正缺的是逻辑，逻辑会产生标准，有了判断标准，混乱会被可预见的选择性所消除。

回到开始的问题，如何开始排版？首先要根据命题信息判断排版的风格，而不能根据自己的喜好或流行来臆断。要了解排版有哪些风格，这些风格有什么特点，有哪些代表作品、代表人物、理论著作，如何形成、何时形成等。再根据风格进行更细致的选择，比如字体、网格等。最后再考虑标点、字距等排版功能技术的规定部分。在这个逻辑中，建立一个风格系统是最重要的环节，这也是目前排版书籍所欠缺的。

排版的风格系统

本书即是从这个角度，把排版分为三大基本风格：古典风格、构成风格、国际风格，并分析其特点、演变、网格等。在风格代表人物的选择上，既要有此风格出色的作品，也要有此风格的理论著作。一个排版风格的形成离不开理论专著的支撑，这也是成为基本风格的前提。同时理论专著除了内容，其本身的版式设计也是了解这个风格的绝好途径。三大基本风格之外的其他风格，都并未有系统的理论专著，而且几乎都衍生于三大基本风格。

本书更像是排版奥义的目录，会对风格的产生、继承、演变做一个脉络分析，并对其经典专著的基本理论进行概括描述，感兴趣的读者，再去有选择地按图索骥，拓展阅读专著的详细内容，进入更细分风格的子宇宙。有了排版的逻辑线与风格系统，之前获得的零散的知识点就会各自归位，变得有条理、系统起来。难点在于大部分理论是英文著作，仅少数专著有中文译本，通读需要大量的连续时间。但这是分水岭，要么安定地研读，到达更高层级，要么永远汲汲于缺少的那一本排版书。

排版的网格系统

网格是很多设计师的关注点,关于网格的书籍也非常多,但仍有很多设计师感到困惑。难点在于:网格制作的着手点;制作过程中的数学关系;基本网格后的无限变化。本书会从这几点详细介绍网格系统。

每种风格都有专属的网格语言,本书根据不同风格、中西文分别介绍网格的使用方法。随着网格制作的深入,数学关系看上去会变得复杂,一定不能中途放弃,一步步理清逻辑与原因,最后才能豁然贯通。

左佐,北京,2018.10

古典风格

古典风格是一种基于传统书籍排印的形式。早于包豪斯风格，特点是常用衬线字体，对称居中布局，正文不分栏等。它不会很前卫，但也从不过时，不会有惊艳带来的侵犯性的瞬间美感，而是一种平静但连绵散发的、需持续关注才会察觉的累积性美感，也正因这种累积性，其随时间推移越久，味道会越浓。

古典风格，不同于传统的中世纪手抄本风格，也不是文艺复兴的人文风格，或是工艺美术运动的自然装饰风格，但有一部分是来源于这些偏手工作坊的传统风格，另一部分则是现代工业的审美。

1931 年 Eric Gill 写了一本书 *An Essay on Typography*（Fig.1），可以看作古典风格的发端。其中"Lettering"一节非常值得阅读，讨论了西文字母的演变历史、字母应用时的忌讳等。在创造上，人都有向好的心（Good Will），但向好的心并不能产出好的事物，设计师还需要好的品位（Good Sense）。

---

1. *An Essay on Typography*，Eric Gill 著，Penguin Books，2013 年版。此书有中文译本。排版必读经典。

2. *An Essay on Typography*，Eric Gill 著，Penguin Books，2013 年版，第 32–33 页。

Classical style

1

ERIC GILL

An essay on
TYPO
GRA
PHY

'Written with clarity, humility and a touch of humour . . . timeless and absorbing'
PAUL RAND, *THE NEW YORK TIMES*

2

32   An Essay on Typography

cause, for him, nothing else was letters; so, in the fifteenth century, when the written was the most common and influential form of lettering, the position is reversed, & the letter-cutter copies the scribe — the stone inscription is imitation pen-writing (with such inevitable small modifications as, in stone, cannot be avoided), whereas in the fourth century the written book was an imitation of the stone inscription (with such small modifications as the pen makes inevitable).

¶ Apart from technical and economic influences the matter is complicated by the differences of individual temperaments and mentalities. Moreover, the physical and spiritual ferment which closed the fifteenth century was accompanied by a revival of interest in and enthusiasm for the things of ancient Greece and Rome, and for the earlier rounder and more legible writing of the ninth & tenth centuries. Nevertheless the first printers were no more the inventors of new letter forms than any other craftsmen had been. The first printed books were simply typographic imitations of pen writing, just as were fifteenth century inscriptions in stone (see fig. 4).

¶ Letters are letters — A is A and B is B — and what we call a gothic A was for Pynson simply A. Print-

Lettering   33

ing started in northern Europe, where the gothic forms were the norm. But the centre of culture was

𝖆𝖇𝖈𝖉𝖊𝖋𝖌𝖍𝖎𝖏𝖐𝖑𝖒𝖓𝖔𝖕
𝖖𝖗𝖘𝖙𝖚𝖛𝖜𝖝𝖞𝖟

(Figure 4: Caslon's Black Letter. This type, like that of Gutenberg, Caxton, &c., was cut in imitation of fifteenth century northern European handwriting. But though the original was handwriting it was for the first printers simply lettering — the only lettering with which they were familiar, book-lettering.)

A B C D E F G H I J K L M
N O P Q R S T U V W X Y Z
a b c d e f g h i j k l m n o p
q r s t u v w x y z

(Figure 5: the Subiaco type. This modern version, cut for the Ashendene Press, London, of the type of Sweynheim and Pannartz, 1465, shows the change in style caused by Italian influence.)

not in the North. German printers moved to the South. The influence of Italian letter forms may be d

古典风格的代表人物是大师 Jan Tschichold，其认为 "Perfect typography is more a science than art"，也正因其技术大于艺术，所以在排印中规则远远多于个人表现。Fig. 3 是其著作 *Treasury of Alphabets and Lettering*，其中收录了从古至今的经典字体样本，介绍了高品位的排印规则。

比如使用大写字母时字间距必须人为调整，比如一个不包含开放字母（如 A、J、L、P、T、V）的单词"HUNDRED"，字间距不能太紧，否则人眼看到的是所有字母的负空间，而非字母的轮廓，导致阅读体验差且不美观。比如大写的词间距，以刚好能放置一个同样字间距的大写"I"为佳。

HUNDRED　HUNDRED　MODERN DESIGN

比如大写字母的行间距，以等于字母高度为佳，也可以是字母高度的 1.5 倍或 2 倍。而且行间距不能小于词间距，否则词间距在横向上看上去像一个孔，影响阅读且不美观。

---

3. *Treasury of Alphabets and Lettering*，Jan Tschichold 著，Van Nostrand Reinhold，1966 年版。排版必读经典。

4. *Treasury of Alphabets and Lettering*，Jan Tschichold 著，Van Nostrand Reinhold，1966 年版，第 30–31 页。

3

4

15

Jan Tschichold 的经历非常有意思，早期在莱比锡受传统排印教育，比如罗马字母表、手抄本、字母书写历史、古老的字体样本等。通过 Edward Johnston 的 *Writing & Illuminating & Lettering* 一书自学了西文书法。1923 年他去看魏玛包豪斯展，被 Nagy 设计的不对称排版的展览手册吸引，之后他又见到了至上主义者 El Lissitzky，在他们的影响下 Jan Tschichold 进入了 New Typography 时期，成为构成风格的先驱之一。

1933 年因其倡导俄罗斯构成主义的新风格，而被纳粹以"文化布尔什维克"为由逮捕。被解救后，他移居瑞士，从事书籍设计。期间他发现构成风格并不为大部分委托人所认同，而且很多学术、人文类的书籍也不适合。在长篇文字排印的易读性上，经典的衬线体远大于无衬线体，如 Garamond、Janson、Baskerville、Bell 等。加之早期的教育根源，之后完全转向古典风格。

Jan Tschichold 对书籍扉页格外关注，认为它是全书的基调，也是读者打开书的第一印象。有两本书至关重要，*Practical Manual of French Typography*（M. Brun, 1825）和 *Handbook of Typography*（Henri Fournier, 1825）。他通过研读两本书中的样例，得到了关于词间距、字体使用、字体尺寸、装饰、规则等各种古典排印细节的奥义。Fig.5 是其著作 *Designing Books*，收录了书籍排印的规则，以及 48 件古典风格作品。

---

5. *Designing Books*，Jan Tschichold 著，Wittenborn/Schulz，1951 年版，扉页。排版必读经典。

6. *Designing Books*，Jan Tschichold 著，Wittenborn/Schulz，1951 年版，Plates 52–53。

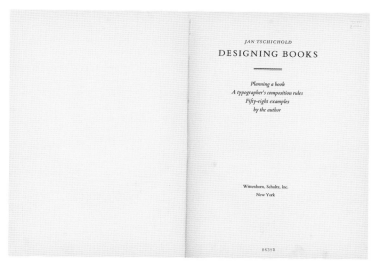

5

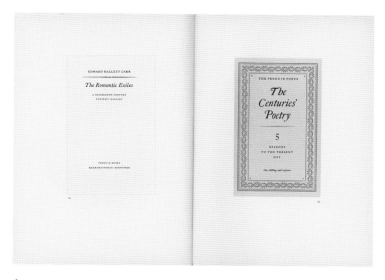

6

1947 年他受邀去英国的企鹅出版社工作，开启了他古典风格的黄金时代。企鹅出于对品牌基因的继承，只允许在原有风格上的改进，Jan Tschichold 运用之前所学，制定了一整套排印规范。开本使用黄金比例的 111mm×181mm；保留原来三分式的构成布局；字体只选择 Monotype 的 Gill Sans，粗体大写用于书名，细体大写用于作者名；在书名与作者名之间加入成为其代表性特点的红色短横线；以颜色区分不同的题材，曾有一个梗说只要黑色的 Gill Sans 出现在深绿色底上，就是典型的英伦风；内文字体上 Garamond、Caslon Old Face 用于乐谱，Bembo 用于莎士比亚系列，优雅纤长的 Bell 用于诗集。

Fig.7 是其改进后的封面设计，构成风格与古典风格相结合，出现了源于传统古籍的经典红色短横线。Fig.8 是 1949 年钱君陶先生设计的《西洋美术史》封面，横分变竖分，红色变绿色。Fig.9 同样是类似风格的版式，中文竖排，可感受到当时中国在设计上是很前沿的。

关于 Jan Tschichold 的经历、作品，可以参阅书籍 *Jan Tschichold: Master Typographer*，或者日本杂志《idea アイデア》的 Jan Tschichold 特辑，321 期（2007.3）。

---

7. *The Woman of Rome*，Alberto Moravia 著，Penguin Books，1956 年版。
8.《西洋美术史》，钱君陶著，永祥印书馆，1949 年版。
9.《中国现代革命运动史》，中国历史研究会编，苏南新华书店，1949 年版。

Classical style

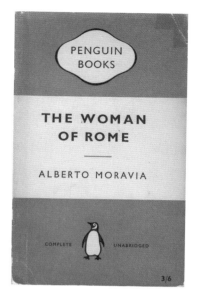

7

8

9

在正文的排版规则（西文语系）下，所有正文的词间距都要尽量紧凑，以字母"i"的厚度为准。为避免大词间距，单词可以自由地断开。正文每行以 8~12 个单词为佳。标题下的第一段不需要首行缩进，其他段落的段首必须缩进 1em（字号大小）。如果一章被分为没有标题的若干回，则每一回结束时需空一行且加入"*"以示结束。章的标题首字母大写, 回的标题用小型大写字母，节的标题用意大利体，左侧页眉为大型小写的书名，右侧页眉为大型小写的章名。意大利体用于着重、外来语、书刊名、演出名。

古典变高数字与现代定高数字不能混用。正文中小于 100 的数字用字母表示，但是年龄、日期等区间数字需使用尽可能少的数字, 中间用 1en（字号的一半）的短横连接，如 1946–7。用（）表示解释和插话，用 [ ] 表示注释。如果每页注解不超过一个，则可使用"*"标注；否则需使用上标数字。页脚的注解需与正文空一行，字号大小比正文小 2pt，行间距与正文相同，首行缩进与正文相同。页码字体需与正文字体相同，使用大小相同的阿拉伯数字。

Fig.10 是 Jane Tschichold 的著作 *The Form of the Book*，是关于书籍装帧的随笔的合集，比如在关于行间距一节，他认为行间距并没有固定的标准，但有几条参考：左对齐的正文，右边越参差不齐，需要的行间距越大，以补偿参差不齐对行带来的视觉干扰；行间距视觉上不能小于词间距，否则词间距的空白会导致串行；每行的单词越多需要的行间距越大。

---

10. *The Form of the Book*，Jan Tschichold 著，Hartley & Marks，1991 年版。排版必读经典。

11. *The Form of the Book*，Jan Tschichold 著，Hartley & Marks，1991 年版，第 92–93 页。正文字体是 Jan Tschichold 设计的 Linotype Sabon。

Classical style

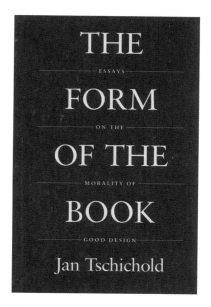

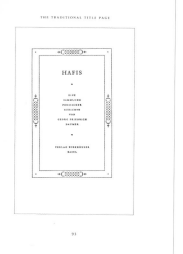

他研究了很多中世纪的手抄本，确信在这些作坊中存在某种排印奥义，像排印圣经的 Gutenberg 也是按此奥义排印，只是未公开于世。这些手抄本开本比例大部分是 2:3，接近黄金比例，取自斐波那契数列，21:34 是现在更精确的黄金比例设定。

在书籍 *The Form of The Book* 中收录了这些手抄本版心设定的几种殊途同归的方式。第一种（Fig.12）页面横竖 9 等分，然后横竖各取 6 份，得到版心；第二种（Fig.13）把页面的宽度作为版心的高度，然后页面高度减去宽度后 9 等分，页边距各取 2:3:4:6 份。

Fig.14 最具延展性，也是中世纪时不借助测量就可以设置版心的秘诀。Fig.15 即是 Fig.14 的延展，在单页高度的 1/3 点延伸横线，得到横线与双页的对角线的交叉点，连接两点得到一个矩形，连接矩形右上点与左边高度的 1/2 点，得到的斜线与单页的对角线产生一个交叉点，由此交叉点横向延伸，得到与双页对角线的交叉点，此交叉点向下延伸与单页的对角线产生交叉点，连接三个交叉点得到版心。同理 Fig.16 在单页高度的 1/8 点延伸横线，得到另一组版心。

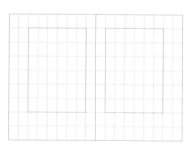

12

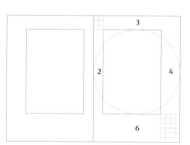

13

Classical style

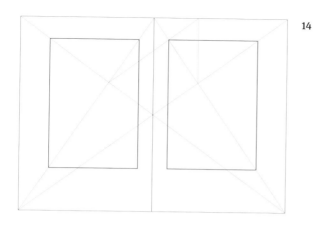

14

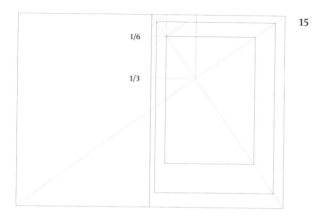

15

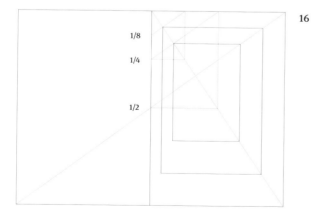

16

西文语境下，古典风格的扉页，是整本书的基调，是读者打开书籍的第一印象。扉页上一般有书名、作者名，也可能有一句解说类话语，位于版心上部；出版者名称、所在地、出版日期等，位于版心下部；也有少数包含一个出版商或作者的标志，出现在版心中部。右页是按古典风格设计的扉页。

对于古典风格来说，扉页的标题必须居中于版心而非页面。扉页字体必须与正文同源，比如正文使用 Old-style 风格的 Garamond，可选择 Garamond、Bembo、Caslon，但如果选择意大利文艺复兴时期的人文风格的 Centaur，或者过渡期的 Baskerville 就会不协调。扉页字体需是同一种，相同粗细，不能使用粗体，字号控制在三种以内。上部书名字号至少是正文的三倍大小；下部的出版信息字号不得大于上部书名信息的最小字号。小写与意大利体不允许调整字间距，大写与小型大写必须调整字间距。数字必须以字母形式出现，年份除外。

上部主行太长断为两行为佳。下部最宽行不得超过上部主行，若超过断为两行为佳。如果上部首行是较宽的主行，则需下移一些；如果下部首行是较宽的主行，则需上移一些。上部首行比主行短，且字号小，是可遇不可求的最佳方式。

---

* 右页是以古典风格的规则设计的扉页。
西文标题：字号：26pt，行距：42pt
西文副标题：字号：16pt，行距：24pt
出版信息：字号：12pt，行距：16pt
中文标题：字号：20pt，行距 42pt
中文作者：字号：11pt，行距：42pt
中文字号以视觉上与西文平衡为准。

左佐 编著

排版的风格
TYPOGRAPHIC STYLE

—

CONSTRUCTIONAL STYLE
INTERNATIONAL STYLE
CLASSICAL STYLE
GRID SYSTEM

—

PUBLISHING HOUSE
OF ELECTRONICS INDUSTRY
BEIJING

排版的风格
TYPOGRAPHIC STYLE

左佐 编著

—

CONSTRUCTIONAL STYLE
INTERNATIONAL STYLE
CLASSICAL STYLE
GRID SYSTEM

—

PUBLISHING HOUSE OF ELECTRONICS INDUSTRY
BEIJING

文字居中页面对齐，并未居中版心；上部标题字号过小；
下部出版信息行宽大于上部书名；上部书名信息偏下。

左佐 编著

排版的风格
TYPOGRAPHIC STYLE

—

CONSTRUCTIONAL STYLE
INTERNATIONAL STYLE
CLASSICAL STYLE
GRID SYSTEM

—

PUBLISHING HOUSE
OF ELECTRONICS INDUSTRY
BEIJING

文字调整为居中版心；上部标题字号调整为正文的三倍，断为两行；下部出版信息行宽过大，断为两行；上部书名信息上移。

瑞士设计师 Jost Hochuli，其风格也横跨构成风格、国际风格、古典风格，有趣的是他年轻时是实验性的构成风格，中年则是理性的国际风格，再后来趋近于优雅的古典风格，可能与当时的流行环境有关，但细想也与这三种风格的特点类似。他最喜欢用的西文字体，是 Old-style 类的 Lexicon 和 Humanist 类的 Trinité。Fig.19 是其著作 *Detail in Typography*，专门收录了更为系统的西文字间距、词间距、行间距的规则。正文字体用 Adobe Minion——文艺复兴晚期的古典风格，标题用 Futura Bold——包豪斯时期的构成风格，两种风格搭配使用竟然有种独特的美感。本书正文与标题也受此启发。

Fig.20 是其另一本古典风格著作 *Modern Typography*，通常认为现代排版始于包豪斯时期或者 20 世纪五六十年代的国际风格，但此书却认为是 1700 年左右，因为此时 Typography 与 Printing 开始分工，并且出现了字体排印样本来规范排版。其中还有大量关于 New traditionalism、New typography、Swiss、typography、Modernity after modernism 等的论述，观点很有启发性。此书全书使用荷兰风格的 Arnhem 字体，在 Erik Spiekermann 最喜欢的字体中排名第五，由第一代数字时代设计师 Fred Smeijers 设计。

---

19. *Detail in Typography*，Jost Hochuli 著，Hyphen Press，2008 年版。排版必读经典。

20. *Modern Typography*，Robin Kinross 著，Hyphen Press，2010 年版。

Classical style

19

Jost Hochuli
Detail in typography
Letters, letterspacing, words, wordspacing, lines, linespacing, columns

New edition

Detail in typography

A concise yet rich discussion of all the small things that enhance the legibility of texts

20

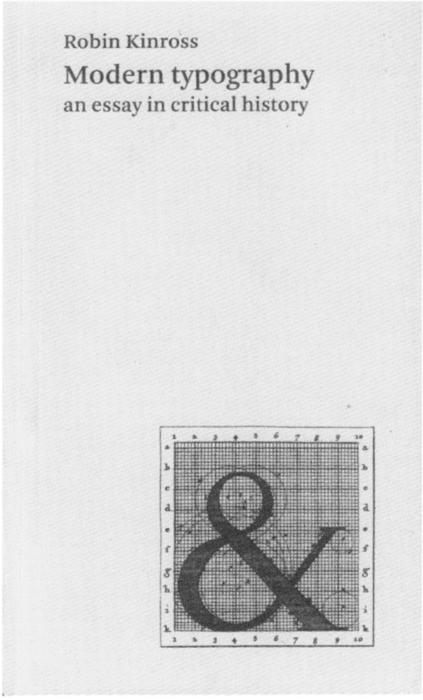

Robin Kinross
Modern typography
an essay in critical history

古典风格的著作着眼点在排版的细节、规则上，不像构成风格有强烈的表达欲，也不像国际风格那样富有理性的秩序感，但古典风格是所有风格的基础。纵观 Typography 的大师，早年均是从古典风格学起，再形成其他风格的。

对于细节规则记叙最为详尽的莫过于 *The Elements of Typographic Style*（Fig.21），作者 Robert Bringhurst 是一名加拿大诗人，其单独的平面类作品很少，大部分是对古典著作和 Typography 历史人物的研究书籍。Old-style 类的 Palatino 是其最常用的字体。

此书通过对传统典籍的研究，得出了很多基于黄金比例、正方形、正五边形、正六边形、圆形的开本比例。同时为了更广的应用，从斐波那契数列延伸出了三组更丰富的数列。a 是黄金比例数列，任意数字都是前面两个数字的和；b 是 a 乘以 2 后的数列；c 是同规则下的另一个数列；d 是 a 与 b 的结合。

a. 1 2 3 5 8 13 21 34 55 89…
b. 2 4 6 10 16 26 42 68 110…
c. 1 3 4 7 11 18 29 47 76…
d. 6 8 10 13 16 21 26 34 42 55 68…

---

21. *The Elements of Typographic Style*，Robert Bringhurst 著，Hartley & Marks，2012 年版。排版必读经典。

22. *The Elements of Typographic Style*，Robert Bringhurst 著，Hartley & Marks，2012 年版，第 186–187 页。

Classical style

21

An early example was Erik van Blokland and Just van Rossum's typeface Beowolf. In its first experimental version (1990), this face relied on the output device to create truly random perturbations from a single set of letterforms. Though it would not work on all systems, and the evolving hardware and software quickly passed it by, it remains an important landmark in the effort to teach computers what typography really entails.

The Hundred-Thousand Character Alphabet

eeeeeeeee eeeeeeeee eeeeeeeee

Beowolf (FontShop, 1990) is at root a statuesque text roman drawn by Erik van Blokland. The letterforms were sent to the output device through a subroutine, devised by Just van Rossum, that provoked distortions of each letter within predetermined limits in unpredetermined ways. Three degrees of randomization – sampled above – were available. Within the specified limits, every letter was a surprise.

Those are two options: manual and random substitution. There is still the third: building glyph-selection rules into the font itself. This gives *predictable* variation. It is now the reigning method for achieving typographic variation, because this is the method built into the OpenType specification.

For now, therefore, the goal of pleasing randomness – constrained but unplanned variation – goes begging in computerized typography. Is it worth pursuing? Communication requires control, just as life requires control – but it also requires a contest beyond its control. Unpremeditated grace is as crucial to the liveliness of the page as it is to the liveliness of the garden.

9.2 THE SUBSTANCE OF THE FONT

Type metal is typically 80% to 85% lead, 13% to 30% antimony and 3% to 10% tin. Some founders also like to add a trace of copper.

Within the tiny confraternity of metal typefounders and letterpress printers there is a subtribe that can argue day and night about recipes for type metal. In such a company, the question of whether to add or subtract five per cent of tin or antimony, or one per cent of copper, can lead to a long and heated exchange. In the community of digital founders and programmers, there is a corresponding subtribe capable of arguing till death about the merits of one digital format versus another.

Between 1980 and 2000, many digital font formats were introduced. Each one's sponsors claimed their product was superior to its predecessors, and sometimes they had grounds to make such claims. In every case, however, it has turned out that what genuinely matters is not the format used so much as the level of hands-on workmanship, good sense and attention to detail. In metal and digital founding alike, the standard is set by the human who does the work, not by the recipe or the brand name of the tools.

Bitmapped fonts came into use in the 1970s. Fonts of this sort are defined by simple addition and subtraction: *this pixel on, that pixel off, these pixels on, those pixels off*. After 1984, with the introduction of the PostScript language, bitmapped printer fonts rapidly gave way to fonts defined as scalable outlines. A decade later came TrueType, which differs from PostScript in several respects. PostScript and TrueType take quite different approaches to hinting (that is, they have different ways of addressing the problems caused by inadequate resolution), and their descriptive mathematics are different. Both interpret letterforms in terms of Bézier splines (in other words, they rely on algebraic techniques developed by Pierre Bézier and Paul de Casteljau in France in the 1960s and 1970s) – but PostScript splines are cubic, while TrueType's are quadratic.

In mechanics, a spline is a flexible strip that will bend under tension. Boatbuilders and furniture makers use them for laying out curves. In mathematics, a spline is a curve that behaves as if it had fiber: sturdy enough to hold itself up yet limber enough to straighten and bend, stretch and retract.

Think of a curve as a tensile line, bent by means of a lever attached to each end. Such a curve can be mathematically defined in terms of four points. Two of these are the *endpoints* of the curve. The other two, at the far ends of the levers, are known as the curve's *control points*. If the two imaginary levers can be controlled independently, then cubic equations [as for example, $f(t) = (1-t)^3 z_1 + 3t(1-t)^2 c_1 + 3t^2(1-t) c_2 + t^3 z_2$] will be required to describe the curve, and it is called a cubic spline.

The levers themselves *are not part of the curve*, and the control points are usually off the curve. In the simplest case, however, these imaginary levers have a length of zero. Then the control points and the endpoints coincide, and the curve is a straight line.

One way to simplify a cubic spline is to tie the levers together so that both control points coincide (or so that one control point has a fixed relation to the other). If that is done, the mathematical description can be simplified from cubic to quadratic [along the lines, $f(t) = (1-t)^2 z_1 + 2t(1-t) c + t^2 z_2$].

There are some other complications. A cubic spline, for instance, can have additional anchors or points of inflection; a

A simple cubic spline, above, and the same curve (more or less) reconstructed below as two quadratic splines.

22

日本设计师白井敬尚是东方的古典风格代表，他翻译了大师 Jan Tschichold 的很多著作，把古典风格的规则元素应用于汉字、假名，得到具有东方韵味的古典风格，主张"极致后的收敛""熟谙规则后的逃脱"。他理论性的专著较少，Fig.23 是他的个人作品集《白井敬尚》。

但作为日本杂志《idea アイデア》的艺术总监，他在大量的杂志中实践其理论，对中文排印非常具有启发性。特别是第 377 期（2017.4）收录的 Typographic Composition，扉页沿用经典红色短横线、居中对称、红黑两色，内页则是多栏排布，衬线与无衬线字体混用，完美诠释了日文语境下的古典风格与国际风格的融合。

另推荐三期《idea アイデア》作为拓展阅读，310 期（2005.5）、314 期（2006.1）、331 期（2008.11），其中 314 期收录的一篇 Typography Review 中的标题形式也是典型的古典风格，红色的"{ }"借用 Jan Tschichold 为其字体 Sabon 设计的字体手册的封面，或许古典风格更多的是一种继承感。

23.《白井敬尚》，白井敬尚著，ggg Books，2017 年版。
24.《idea アイデア》，2005.5，321 期，诚文堂新光社。
25.《idea アイデア》，2008.11，331 期，诚文堂新光社。

YOSHIHISA SHIRAI
白井敬尚

白井さんは、解決でも表現でもない、デザインによる批評として、著者、編集者と絡み合うプログレッシブな音を響かせてきた。　室賀清徳

古典风格的开本比例常取自基本的几何图形,偏爱瘦长的黄金比例开本,与国际风格偏爱正方形开本形成对比。

```
341-2400                    INSTITUTE FOR
                           ADVANCED STUDY


              ALBERT EINSTEIN
                    PROFESSOR



   112 MERCER ST.
   PRINCETON UNIVERSITY          PRINCETON, NJ.
```

古典风格在中文语境下也有很多应用，只是年代久远，不太为人关注。比如，民国时的很多名片、信封、包装的版式都属于古典风格。还有一类被称为"红色美学"，是很高级的一种古典风格，排版方式极度简洁、极具威严感。下面是笔者收集的一些资料，供读者参阅。

26.《中共中央文件》,中共山西省委办公厅,1965年。

27

27.《参观毛主席故乡留念》日记簿,长沙印刷厂,1967年。

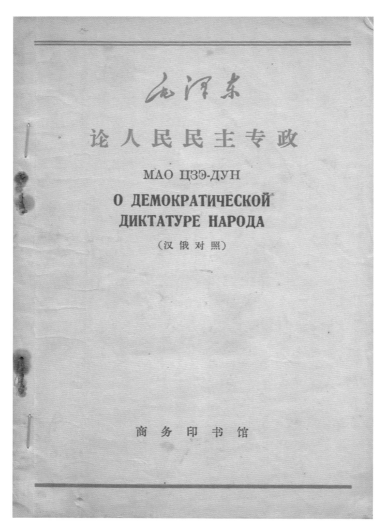

28.《论人民民主专政》，毛泽东著，商务印书馆，1966年。

29

29.《马克思、恩格斯、列宁、斯大林、毛泽东著作和党中央负责同志言论摘录》，中共辽宁省委宣传部编，辽宁人民出版社，1960年。

30. 民国时期的私立金陵大学图书馆信封。

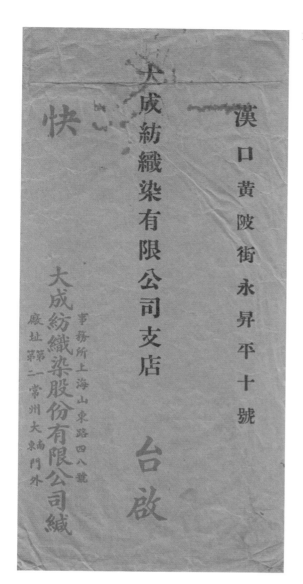

31. 民国时期的大成纺织染股份有限公司信封。

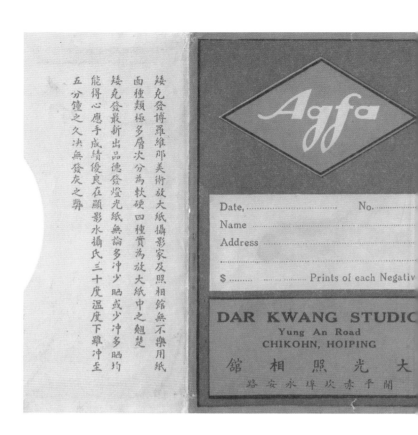

古典风格中中西文混排时,同样一段信息,一般西文行宽大于中文,所以中文会增加字间距来适配,同时字间距会根据中文字数的多少而变化,形成中文特有的排版纹理。比如 Fig.32 中的中西文混排。

32. 民国时期的矮克发（Agfa）包装盒。

古典风格需要对文字本身很敏感，间距、大小、位置都一点点细腻地微调，而这种变化大部分不会被大众感知，至多会形容为"舒服"，再难有其他；但对于高品位的排版者，这种变化带来的视觉感受则被无限放大，由此带来的美的感知也是无穷尽的。

Fig.33 是笔者为《设计师的自我修养（修订版）》设计的封面，即是受 Jan Tschichold 与白井敬尚的影响，红色短横、居中对称、衬线字体，是较为典型的古典风格。

古典风格并不太适合当下"短时高效"的互联网式审美口味，不过审美从来都是一个轮回，在各种强刺激设计漫天飞舞后，人类趋于麻木的感官会逐渐转向细腻、优雅、恒久的古典风格。

33.《设计师的自我修养（修订版）》,左佐编著,电子工业出版社, 2018 年版。

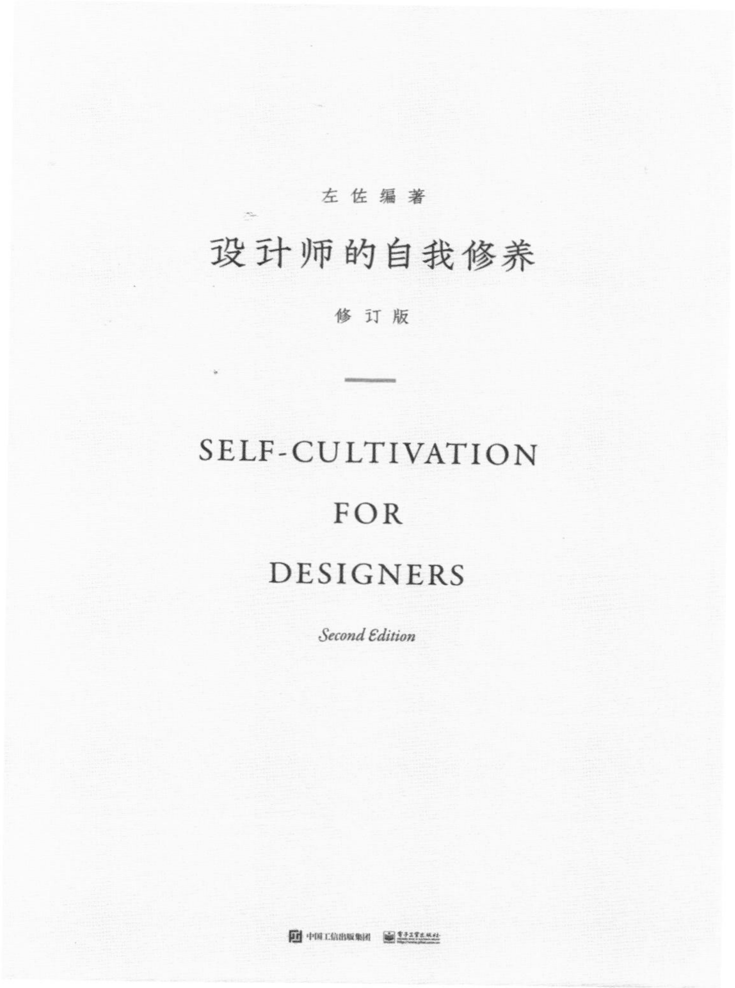

左佐 编著

设计师的自我修养

修订版

—

SELF-CULTIVATION
FOR
DESIGNERS

Second Edition

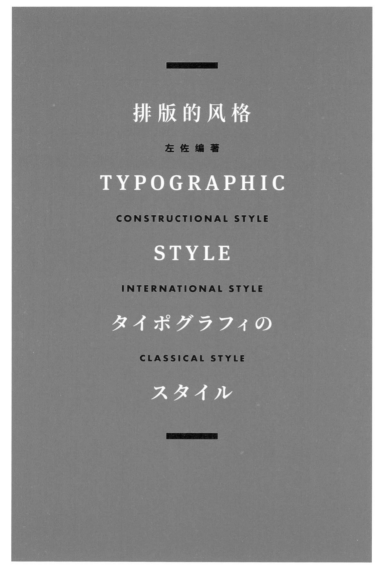

以古典风格与构成风格结合来设计封面。字体选用衬线体 Meta Serif，居中对称；副标题部分使用偏几何的 Futura Bold，英文字间距都经过细调，中文由思源宋修改而来。再配合粗重的横线，古典风格中融入构成风格。

版式借用民国时期名片的竖排居中形式；标题选用衬线体 Garamond；副标题、作者名、出版社名则选择无衬线体 Din，对称分布在四边；中文标题为中宫紧缩的定制字形与西文搭配；再搭配古典风格常用的线框和装饰图案。

构成风格

20 世纪 20 年代受俄罗斯构成主义影响，出现了 New Typography 风格，有两个核心观点：认为无衬线体是这个时代的字体；中心对称是旧时代的束缚，只有非对称审美才是解脱。此种风格，以及由此风格延续而来的风格，称为构成风格。构成风格出现于包豪斯时期，所以包豪斯在排印方面有很多造诣颇深的大师，只是少有人关注，比如 Moholy Nagy 和之后的 Herber Bayer，但因年代久远，两人传下来的作品很少。

构成风格的排版有一个方形法则，即排版中的文字、线条、空白会刻意或无意地组合成不闭合的方形（Fig.37/39），这与当时非常流行的方形崇拜有关。方形法则可以用完形心理学中的"闭合"来解释，眼睛在观看时，会把各部分组合起来，使之成为更易于理解的统一体。这里不闭合的部分图形，通过眼睛闭合为统一的正方形，从而带来美感。

杂乱无序的排版或图形会让眼睛无从判断落点，无法形成整体认知，就会让读者因不知如何阅读而产生焦虑，这也说明韵律是眼睛喜欢的，它会让眼睛按一定的规律去阅读，进而产生安定感、成就感、愉悦感。

---

36. Moholy Nagy 于 1923 年设计的信头。
选自《包豪斯》，Jeannine Fiedler 等编著，浙江人民美术出版社，2013 年版，第 335 页。

38. Herber Bayer 于 1924 年设计的信头。
选自 Think with Type，Ellen Lupton 编著，Princeton Architectural Press，2010 年版，第 162 页。

Constructional style

36

37

38

39

之后这种风格得到一位不世出的天才大师 Jan Tschichold 的延续。之所以说他不世出，因为他既有绝佳作品，又著述极多，而且几乎每本都是经典。日本几乎翻译了他的全部著作，而我们貌似一本都没有。

Fig.40 是 1925 年他设计的封面，信息分三部分，最外圈、中心、右下角，阅读顺序由线条和文字大小来引导。Fig.41 标示出方形法则的结构，最明显的是右下角，为形成方形而特意增加了竖线元素，这也是构成风格的手法之一。

那个时代全世界都在被方形、圆形、三角形等基本图形洗礼，中国也不例外。Fig.42 是民国旧书《英国文学研究》的封面，左下角的英文"MODERN"表明这种风格在当时是非常前卫的。Fig.43 标示出方形法则结构，可以左右对比来看隐藏的方形对视觉的影响。

眼睛在看到线的时候，会沿线移动，所以线可以引导阅读。Fig.42 中的粗实线会让信息被优先获取，而阶梯形会标示获取的顺序。在众多信息中，眼睛很容易失焦，不知道从何开始阅读。当眼睛看到圆形时，一般是静止的，这时适当地使用圆形可以起到降低阅读焦虑的作用。封面中黑色的大圆形，除了平衡空间外，也可以让眼睛一下就定在这个位置，进而从"英"字开始阅读。

---

40. 选自杂志《idea アイデア》，321 期（2007.3），第 37 页。此期为 Jan Tschichold 特辑，排版必读经典。

42.《英国文学研究》，小泉八云著，现代书局，1932 年版。

40

41

42

43

描述一下前面几件作品的细节：红黑色；线、圆等基本图形；标题为粗体，次要信息为对比较明显的细体；不同文字块排列时，会左右交错；文字阅读方向会有纵横两种；字体一般为 Grotesque 类无衬线体等。这会锻炼设计师的观察力与描述力。

而且越早期的作品，隐藏的信息越多，了解得越深入，领悟到的越多，越能领会这种风格的奥义。类似练楷书，必先参篆隶。细节描述也会发现规则，改变其中的若干规则，就会出现风格的演化可能性。

Fig.44 是书籍 *Typographische Gestaltung*（Typographic Design）的封面。此书是 Jan Tschichold 影响力最大的书籍之一，出版于 1935 年。可见粗线变为细线；标题用 Egyptian 类平板衬线体，小型文字用现代衬线体 Bodoni，无衬线体被抛弃，粗犷、建筑、构成的感觉弱化，由构成风格向古典风格过渡。Fig.45 是版式的方形法则结构线。

Fig.46 是民国旧书《雪》的封面，采用汉字语境下的构成风格，元素变为贯穿页面的红色粗线，方形法则同样适用。

---

44. 选自 *Jan Tschichold: Master Typographer*，Cees W. de Jong 等编著，Thames & Hudson，2008 年版，第 134 页。

46.《雪》，巴金著，文化生活出版社，1949 年版。

44

45

46

47

Constructional style

在瑞士，设计师 Jost Hochuli 明显继承了构成风格，Fig.48 是他为自己的著作 *Designing Books* 设计的封面，融入了瑞士人的细腻，古典的衬线体被引入，把最初德国、俄罗斯构成风格的粗犷感降低了，但仍能看出是构成风格——红黑两色、线条、空间的分割、阅读的引导、字体的选择、文字的交错形成的隐藏方形。

Fig.50 是他另一本著作 *Printed Matter, Mainly Books* 的封面，把经典的红黑变为红黄蓝绿的基本色，小字是经典的衬线体。他通过改变字体、颜色规则（小字的古典衬线与大字的无衬线）演化出构成风格的另一种可能性。但是方形法则同样适用。

---

48. *Designing Books: Practice and Theory*，Jost Hochuli 著，Hyphen Press，1997 年版。

50. *Printed Matter, Mainly Books*，Jost Hochuli 著，Niggli Verlag，2001 年版。

Constructional style

48

49

50

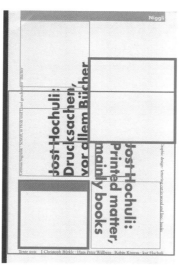
51

日本对构成风格并不太热衷，可能跟其细腻、感性的民族特性有关，瑞士人虽也细腻，但却是理性的细腻，所以有构成风格的土壤。Fig.52 是小泉均的著作 *Handbook of Typography* 的封面，曾在日本引起热议，从中可以看出构成风格的影子，粗实的线、交错的文字块、圆形的标志、红黑配色、黑体等。

Fig.53 是其方形法则结构线，粗线是主要的隐藏方形，符合方形法则的设计大多很美，所以可以作为一个规律来使用，以产生同样美的设计。结构线有时并不完全符合正方形，一个原因是视觉补偿，另一个原因是设计者并未意识到这个规律，大部分是无限接近于正方形。

Fig.54 是 Fig.44 的日文译本《アシンメトリック・タイポグラフィ》（不对称排印）。由西野洋装帧设计，虽然是参考原书的装帧，但仍能感觉到明显的不同，开本、字号大小、字体选择、颜色设定，都散发着一股独特的感性的细腻气质，也许这就是日本能接受的本土化的构成风格吧。

---

52. *Handbook of Typography for Students and Practitioners*，小泉均编著，研究社，2012 年版。排版必读经典。

54.《アシンメトリック・タイポグラフィ》，Jan Tschichold 著，鹿岛出版会，2013 年版。排版必读经典。

Constructional style

52

53

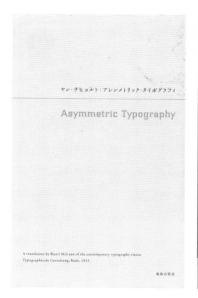

54

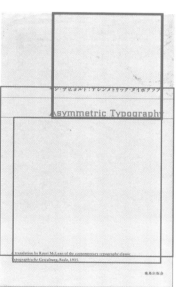

55

文字较多的内页的构成风格。Fig.56 是非常有代表性的作品，1928 年由 Jan Tschichold 设计，不对称的分栏，文字块的交错，线条的粗细、大小、位置，文字的大小、粗细，留白的位置、面积等，都非常值得深究，不管大师是否有意安排，这件作品同样符合方形法则，时常参悟对获得高级排版奥义助益极大。

56. 选自 *Jan Tschichold: Master Typographer*，Cees W. de Jong 等编著，Thames & Hudson，2008 年版，第 101 页。

Fig.58 是典型的构成风格内文，1925 年由 Jan Tschichold 设计。线与文本行并列，粗重的红色线用来隔开信息块，同时让版面不只有文字。对于通篇密密麻麻的文字排版，大部分人都会未读先倦，长时间看类似形式的信息同样会疲劳，而这些线则会将大信息分割为小信息，同时眼睛会看到不同的形式，产生新奇感，起到缓解的作用。

58. 选自 *Meggs' History of Graphic Design*, Philip B. Meggs 等编著, Wiley, 2006 年版, 第 320 页。

Fig.59 是 Jost Hochuli 为书籍 *Detail in Typography* 设计的内页。粗线变细，字体变为衬线，构成的线与古典的字并列，形成独特的美感，是构成风格与古典风格的融合。古典风格在于不分栏，只一栏，衬线做正文；构成风格在于红线、页码、章节名的字体、位置、大小。

The typefaces used
– Adobe Minion Regular
– Adobe Minion Expert Regular
– Adobe Minion Italic
– Futura Bold

a

The typefaces used
– Adobe Minion Regular
– Adobe Minion Expert Regular
– Adobe Minion Italic
– Futura Bold

b

The typefaces used
• Adobe Minion Regular
• Adobe Minion Expert Regular
• Adobe Minion Italic
• Futura Bold

c

The typefaces used
· Adobe Minion Regular
· Adobe Minion Expert Regular
· Adobe Minion Italic
· Futura Bold

d

50 The en dash used for lists. It is separated from the following word by one or more wordspaces. The dash can be replaced by either bold (c) or normal (d) centred points. The style chosen will depend on the typeface and the text; depending on the typographic design, the points or dashes may be aligned under the line above (a, c) or hung out (b, d).

Pablo Casals/Alfred Cortot/Jacques Thibaud
Pablo Casals / Alfred Cortot / Jacques Thibaud

51 A piece of typographical nonsense often encountered: solidus strokes set unspaced – that which belongs together has been pulled apart.

in some European languages to indicate quotations (« ») are set with the points facing outwards, as shown here, in France and Switzerland, but with the points facing inwards in Germany.

Abbreviations in capitals are often too obtrusive, and interfere with the smooth flow of the line. Instead of capitals, they can be set in small caps, but this may look strange. Another possibility – and often the best – is to set the abbreviation a half or one size smaller, depending on the typeface. The weight of the letters will no longer be identical with that of the continuous text, but this is optically hardly noticeable [54]. (See also the discussion of capitals under 'Emphasis', p. 44.)

59. 选自 *Detail in Typography*, Jost Hochuli 著, Hyphen Press, 2008 年版, 第 38–39 页。排版必读经典。

虽说日本民族特性并不适合构成风格，但无印良品是个例外，充满构成感。Fig. 60 是田中一光 1991 年设计的 MUJI 标志，Grotesque 风格的无衬线粗体，中间黑色的粗重线条，线条中的文字是 Helvetica 细体，这是典型的构成风格。

Fig.61 是 MUJI 的标签设计，是对当时日本包装过剩、色彩使用过剩的反制，以一种无设计的形象面世，粗线分隔、红黑配色、无衬线黑体。值得注意的是，文字是居中对称，而非构成风格的不对称排版，从而多了一种古典优雅的味道，这可能是日本特有的柔软的构成风格吧。

常说 Design is history，看多了不同时期的优秀作品会越发认同。早在包豪斯时期，各种手法风格早都已经有了花种，那时的人大部分是艺术与工艺并举，是设计师的最佳形态。之后的经典设计都或多或少从中获取花种，再以各自的时代和地域的审美培育出花。

这也是分析解构经典的意义，一件高级的作品，一定是对过去经典的解构再输出。也许会说这件作品如此叛逆，以前从未有过，但这叛逆恰好是针对经典而言的，是受经典的刺激而向其反方向逃离的行为。这也是当今鲜有大师的原因，一代一代的解构越来越多，后人可输出的前人已基本完成，剩下一些凤毛麟角的部分已经大大稀释了输出作品的影响力。所以未来著名设计师会越来越多，而鲜有大师。

---

60. 选自《田中一光的设计世界》，朱锷主编，中国青年出版社，1998 年版，第 122 页。
61. 选自《無印良品》，無印良品著，朱锷译，广西师范大学出版社，2010 年版，第 155 页。

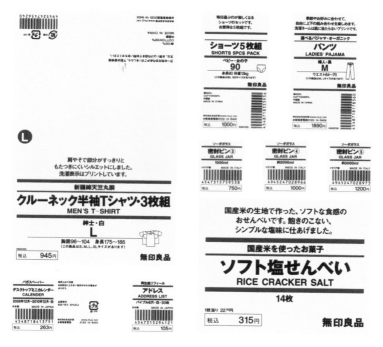

Constructional style

03

Regarding to materials, there is only definition in the philosophy of Feelfel natural.We reject any form of damage to the nature, firmly choose to import recycled materials from Germany to make "felt" products for the life, and use renewable resources to promote sustaina development.
Bearing the concept of "from the nature and back to the nature", each product expressed in the form of natural, which makes the product simple, and through infinite material to embrace the independent spiritual proposition and free life attitude.
"Feelfelt" products are free of animal chemical ingredients and are ideal for many allergies and vegetarians. They a non-toxic and degradable, also safe fo children.

THE MATERIALS

关于材料，「复觉」的哲学里只有一个定义—自然。我们拒绝对自然形式上的破坏，坚定选择德国进口再生材料，制成覆盖生活的「毡」类产品，利用可再生资源，倡导可持续发展。信守一始于自然，归于自然的理念。每一件产品皆通过设计的形式表达，使产品化繁为简，通过材质的无限，容纳独立精神主张和自由生活态度。「复觉」产品不含动物和化学成分，非常适合许多过敏症和素食主义者。非毒，可降解，幼儿亦可安心使用。

构成风格，经过不同的人、不同的时代会自然衍生出一种有继承感的形式，这种有渊源的作品，比无基础的凭空创造更有价值。Fig.62是工作室为"复觉"设计的宣传册内页，复觉的产品是用可降解的毛毡制成的，粗犷、自然，简单的块面感，与构成风格很契合，所以根据实际委托对构成风格进行了解构。

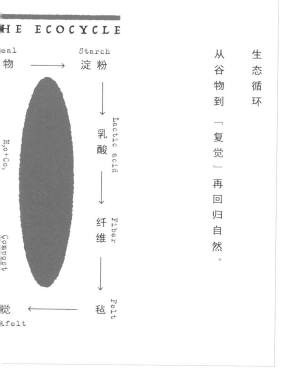

当下中国构成风格的影子较稀有,主要流行的是宝岛风格,偏甜腻小情调。看过一本由方建平和丁凡设计的书籍《杜尚与 / 或 / 在中国》,文中贯穿的红色分割细线,边缘略粗的红色线,简单的一栏文本,没有刻意安排过多的穿插错落,衬线字体的选择,是难得的构成风格与古典风格的融合。其实早在民国时,钱君匋先生就对汉字的构成风格做了很多独特尝试,推荐商务印书馆 1992 年版的《钱君匋装帧艺术》拓展阅读。

03

Regarding to materials, there is only o
definition in the philosophy of Feelfe
natural.We reject any form of damage to
the nature, firmly choose to import
recycled materials from Germany to make
"felt" products for the life, and use
renewable resources to promote sustaina
development.
Bearing the concept of "from the nature
and back to the nature", each product i
expressed in the form of natural, which
makes the product simple, and through t
infinite material to embrace the
independent spiritual proposition and t
free life attitude.
"Feelfelt" products are free of animal
chemical ingredients and are ideal for
many allergies and vegetarians. They ar
non-toxic and degradable, also safe for
children.

THE MATERIALS

关于材料，"复觉"的哲学里只有一个定义——自然。我们拒绝对自然任何形式上的破坏，坚定选择德国进口再生材料，制成覆盖生活的"毡"类产品，利用可再生资源，倡导可持续发展。

信守"一始于自然，归于自然"的设计材质的形理念，每一件产品皆通过去通过材质的形式表达，使产品化繁为简，和自由生活态度的无限，容纳独立精神主张。

"复觉"产品不含动物和化学成分，非常适合许多过敏症和素食主义者。无毒，可降解，幼儿亦可安心使用。

宣传册的方形法则结构线。

Constructional style

4 From nature to "feelfelt", then back to nature.

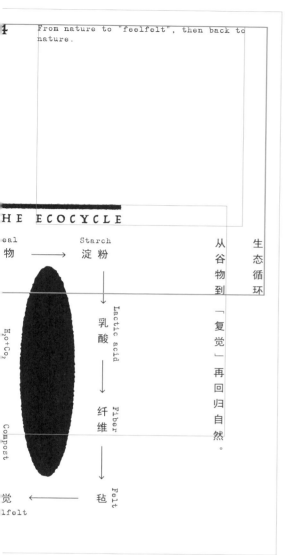

THE ECOCYCLE

Cereal → Starch → Lactic acid → Fiber → Felt → Feelfelt
$H_2O + CO_2$
Compost

生态循环

从谷物到「复觉」再回归自然。

谷物 → 淀粉 → 乳酸 → 纤维 → 毡 → 复觉

排版的
风格

左佐
编著

TYPOGRAPHIC
STYLE

CONSTRUCTIONAL STYLE
INTERNATIONAL STYLE
CLASSICAL STYLE
GRID SYSTEM

用构成风格设计的封面。观察与描述训练多且精准后，总结出的规则，根据实际委托放弃规则，或用新规则代替，就会出现风格的各种可能性。

排版的
风格

左佐
编著

TYPOGRAPHIC
STYLE

CONSTRUCTIONAL STYLE
INTERNATIONAL STYLE
CLASSICAL STYLE
GRID SYSTEM

根据方形原则把结构线极度简化，方便应用方形法则，采用无衬线体 Akzidenz Grotesk，文字块交错，文字、图形、留白形成大小不一的方形。最下方的文字根据实际情况微调，并未遵循结构线，弹性使用规则是作品灵动的来源。

TYPOGRAPHIC STYLE

排 版 的 风 格

左 佐 编 著

Constructional Style
International Style
Classical Style
Grid System

电子工业出版社

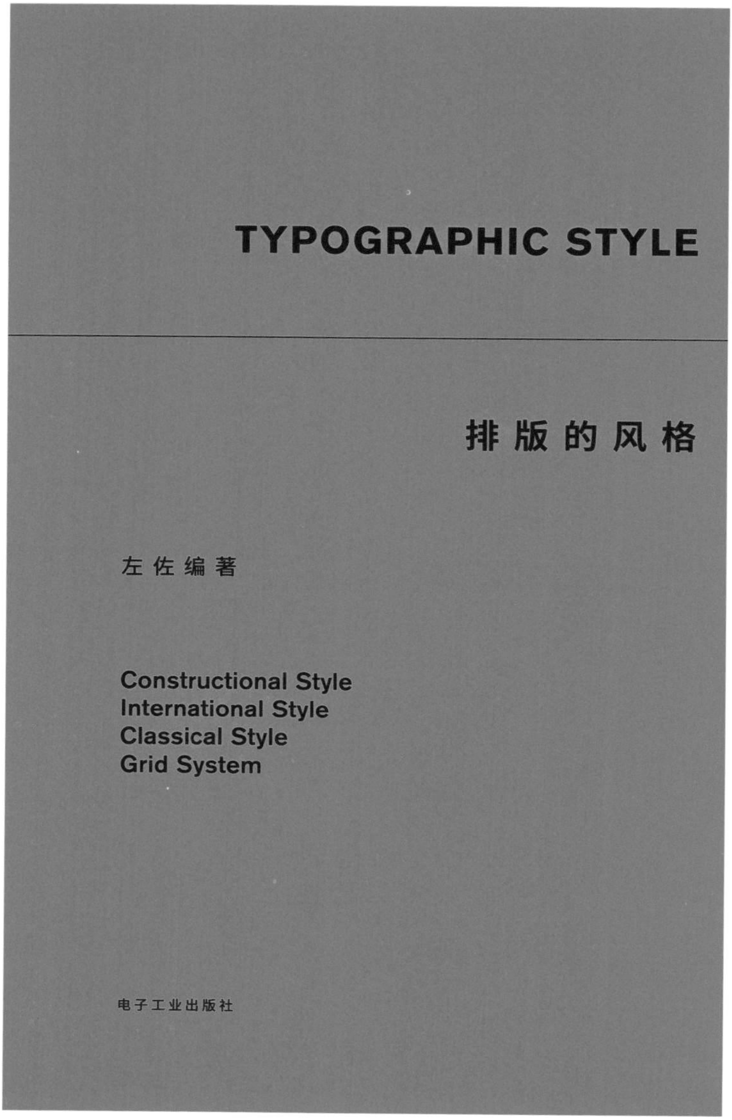

用构成风格设计的另一款封面，封面上的线条变细。

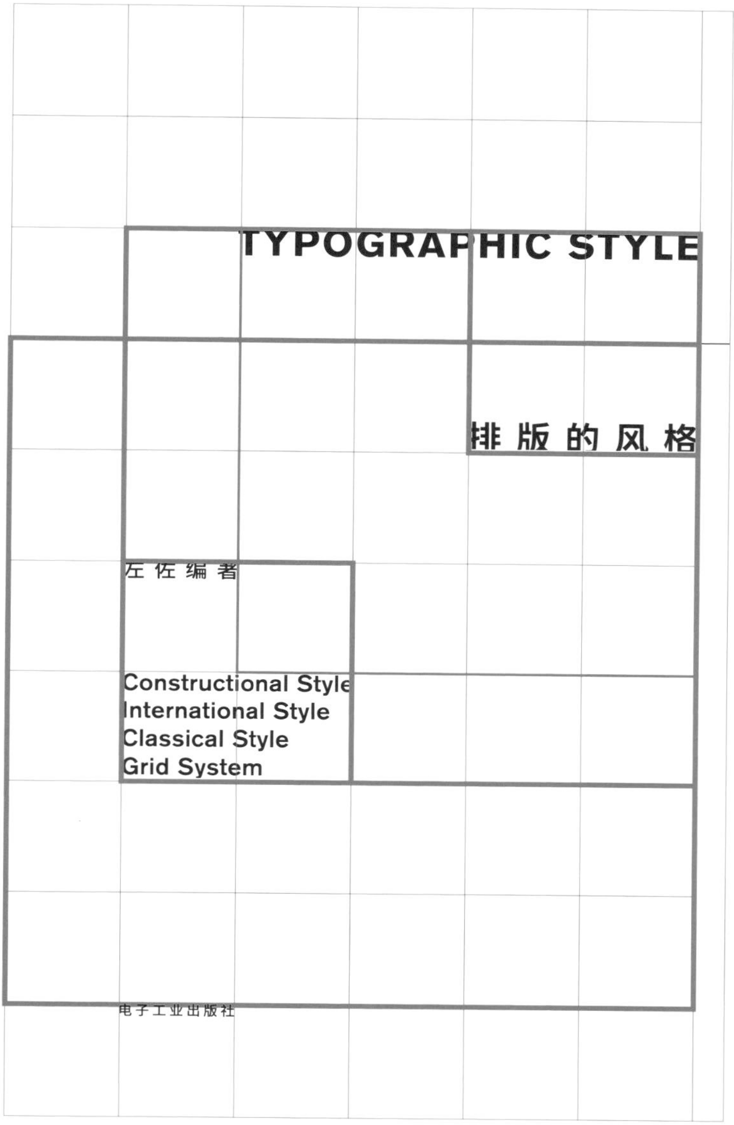

Constructional style

TYPOGRAPHIC STYLE

排版的风格

左佐编著

Constructional Style
International Style
Classical Style
Grid System

电子工业出版社

构成风格的极简结构线，严格适用方形法则。

国际风格

国际风格,也叫瑞士风格,形成于20世纪五六十年代,是构成风格在瑞士的本土化,以平均分布代替不对称交错,细腻的理性代替粗犷的构成感。

Fig.68是通常认为的国际风格网格,正方形开本,可以说任何人都可以制作这种分栏,因为它是平均分布的,不需要过多依赖设计师的品位判断。图文的排布也是如此,是一种有规则可循的变化。比如图片宽度,任何设计师控制都类似,不过是一栏、二栏、三栏、四栏的不同,具有非常大的通用性。即使品位不高的设计师,也容易设计出品位不俗的作品。

但其实这是对国际风格的片面理解,国际风格内部并不像看到的那样统一,可以分为两个派别:苏黎世学派,理性客观,即通常认为的国际风格;巴塞尔学派,感性主观,常被忽视的国际风格。

68

苏黎世学派以苏黎世设计学院（Zurich School of Design）为教学地，以 Josef Müller-Brockmann 为代表，以理性的网格系统为代表理念。

Fig.69 是 Brockmann 的著作 *Grid Systems in Graphic Design*（有中文译本），主张设计应该易懂、客观，具有功能性、具有数学逻辑美感，通过网格系统，图片会精简到几种相同的尺寸，文本会清晰且有逻辑地分出标题、副标题、正文、注解。这种风格在之后兴起的品牌 VI 手册中应用极广。

设计并不是个人观点或感情的自我表达，而是基于商业委托的系统化、清晰化、构成化的客观解决方案。也因此被当时的巴塞尔学派戏称"Swiss square is doing well"，可见其理念上的分歧。

---

69. *Grid Systems in Graphic Design*，Josef Müller-Brockmann 著，Niggli Verlag，2012 年英文版。排版必读经典。

70. *Grid Systems in Graphic Design*，Josef Müller-Brockmann 著，Niggli Verlag，2012 年英文版，第 50–51 页。

International style

巴塞尔学派，以巴塞尔设计学院（Basel School of Design）为教学地，以 Emil Ruder 为代表，以规则内的主观实验为代表理念。Fig.71 是 Emil Ruder 的著作，有中文译本。他主张"From the Inside"，强调人性（Humanity）的部分，发现了排版中"积极的负空间"（Positive Negative Space），鼓励学生进行各种实验，但实验是以当时活字印刷技术的规则为前提的，也因其主观实验性，作品会有无逻辑、不可说、以个人品位安排的部分。

Fig.72 右页是 Emil Ruder 的作品，其中的正方形，在为何使用、为何放在这个位置、为何是这个大小方面并无逻辑，但版面中的文字、空白、图形却有着某种动态的平静感。Fig.74 左页是 EI Lissitzky 的实验作品，其无逻辑、主观的部分与前者相似，不同的是 Emil Ruder 是在规则下的实验，而 EI Lissitzky 则是对已有规则的完全推翻。

还有一本以第三视角介绍他的排版理论的杂志特辑，《idea アイデア》第 333 期（2009.3），推荐拓展阅读。

---

71. *Typogrphie*，Emil Ruder 著，Niggli Verlag，2009 年英文版。排版必读经典。

72. *Typogrphie*，Emil Ruder 著，Niggli Verlag，2009 年英文版 , 第 112–113 页。

International style

71

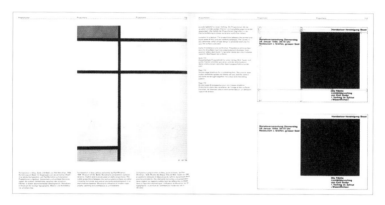

72

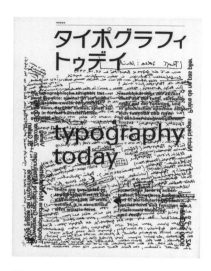

73

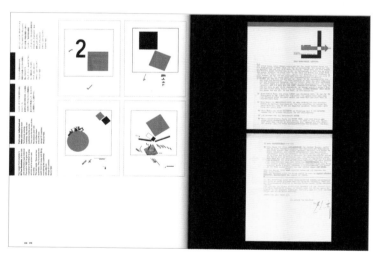

74

73. *Typography Today*,Helmut Schmid 著,诚文堂新光社,2015 年版。此书有中文译本,收录了从构成风格到国际风格的代表设计师的作品、言论,很精彩。排版必读经典。

74. *Typography Today*,Helmut Schmid 著,诚文堂新光社,2015 年版,第 178–179 页。

75

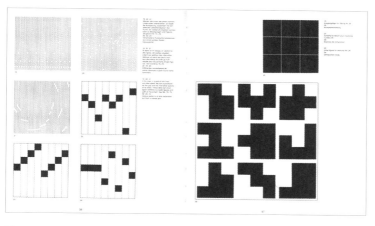

76

International style

75. *Graphic Design Manual, Principles and Practices*，Armin Hofmann 著，Niggli Verlag，2015 年英文版。作者也是巴塞尔学派的代表人物，著作从图形的角度来进行实验，试图发现一种不随科技工具变化而改变的永恒规则。

76. *Graphic Design Manual, Principles and Practices*，Armin Hofmann 著，Niggli Verlag，2015 年英文版，第 112–113 页。

两个派别各有自己的发声杂志，苏黎世学派的代表杂志，是 1958 年由 Müller-Brockmann 参与创立的 *New Graphic Design*，在创立之初 Müller-Brockmann 曾提到请巴塞尔学派的 Emil Ruder 加入，但遭到其他人的反对，因为他们不相信过于主观的巴塞尔设计师。Fig.77 是杂志的复刻版，完整再现了学派的理论及应用作品。

杂志开本为方形，为体现其构成、理性的特点，严格遵循 4×4 网格，文本块窄且长，左右皆强制对齐，段落起始无空格，且段落与段落之间无空行，这样，从视觉上文本块会形成规整的矩形，整体版面极度理性。当然，过于追求形式上的理性，也牺牲了易读性，据说很少有人真正阅读杂志的文字。杂志的撰稿人几乎都是苏黎世学派的瑞士设计师，有很强的排外性。这种过于理性的排版，变化严格遵循网格，是完全可预见的，不断重复后很容易让人感觉乏味无趣。

杂志选录的作品，大部分是图文混排的商业作品（Fig.78），个人实验是极少的，可以看出苏黎世学派的重点在于 Graphic Design。因商业作品实用性大于实验性，所以其风格更偏向于网格。但商业作品的传播性远大于个人实验，所以一般大众理解的瑞士风格，更多的是苏黎世学派的偏网格的风格。

77. *New Graphic Design: 1958–1965*，Lars Müller 编，Lars Müller Publishers，2015 年版，第 2 期。

78. *New Graphic Design: 1958–1965*，Lars Müller 编，Lars Müller Publishers，2015 年版，第 2 期，第 30–31 页。

International style

77

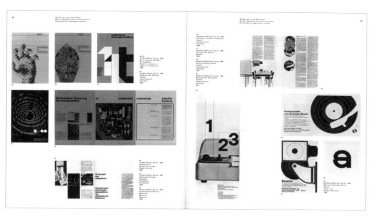

78

巴塞尔学派的发声杂志是 1922 年创办的 *Typografische Monatsblätter*（Typographic Monthly），简称 TM，每年会由一个设计师来设计封面及内页版式，所以不会有严格的网格限制，具有很大的实验性，风格也因人而异，充分展示了主观的优势。Fig.79 是关于此杂志的书籍，以图文的形式介绍了国际风格的先驱们的作品、理念、教学方法、相互关系等。

杂志选录的作品大部分是单纯的 Typography 文字实验（Fig.80），图文混排的商业作品较少，但其拥有很完整的教学方法，在院校中较为盛行，偏学院风，传承上有很大优势。在风格的延续上，苏黎世学派大多是知名的商业设计师，著作以作品集配合注解的形式出现；巴塞尔学派大多是知名的院校教师，著作以教学方法配合应用作品的形式出现。

更抽象地说，应该是应用经验派与实验理论派。理论派在研究文字的各种可能性，而可能性会被经验派应用于商业设计。反观汉字也类似，应用经验的书籍很多，但基础实验理论的研究很稀缺。

---

79. *30 Years of Swiss Typographic Discourse in the Typografische Monatsblätter: TM RSI SGM 1960–90*，Louise Paradis 等编，Lars Müller Publishers，2017 年版。排版必读经典。

80. *30 Years of Swiss Typographic Discourse in the Typografische Monatsblätter: TM RSI SGM 1960–90*，Louise Paradis 等编，Lars Müller Publishers，2017 年版，第 108–109 页。左页是固定开本、固定字体、固定字号、固定文本，在如此限制下，对于排版的练习，仍然有无限的可能性。

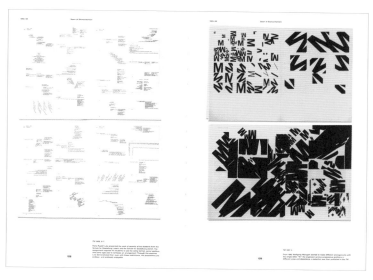

International style

两个学派对于字体的偏好也截然不同，苏黎世学派偏爱 Akzidenz Grotesk 与 Helvetica，*New Graphic Design* 杂志全部使用 Akzidenz Grotesk; 而巴塞尔学派则以 Univers 为代表，*TM* 杂志几乎全部使用 Univers 字体。

Akzidenz Grotesk 中的 Akzidenz 意为商业的（Commercial），1898 年由德国的 Berthold Type Foundry 发行，属于 Grotesque 类无衬线体，用于公共广告、票据、表格等商业印刷，是最早发行的无衬线体之一。

Helvetica 意为瑞士的，1957 年由瑞士的 Haas Type Foundry 发行，是为了与 Akzidenz Grotesk 竞争而开发的同类字体，属于 Neo-grotesque 无衬线体，目前应用较多的是各大品牌的标志字、企业用字。

Univers 意为通用的，1957 年由受雇于法国铸字公司的瑞士设计师 Adrian Frutiger 设计，同样受 Akzidenz Grotesk 为代表的德国无衬线字体影响，属于 Neo-grotesque 无衬线体，目前应用较多的是书籍、环境导视。

---

* 关于 Helvetica 可参阅书籍《字体传奇：影响世界的 Helvetica》，Lars Müller 编，李德庚译，重庆大学出版社，2013 年版。

* 关于 Univers 可参阅书籍 *Adrian Frutiger–Typefaces: The Complete Works*，Adrian Frutiger 等著，Birkhauser，2014 年版。

**abcdefghijklmnopqrstuvwxyz
ABCDEFGHIJKLMN
OPQRSTUVWXYZ
0123456789**

Akzidenz-Grotesk BQ
Medium

**abcdefghijklmnopqrstuvwxyz
ABCDEFGHIJKLMN
OPQRSTUVWXYZ
0123456789**

Helvetica Neue LT Std
66 Medium

**abcdefghijklmnopqrstuvwxyz
ABCDEFGHIJKLMN
OPQRSTUVWXYZ
0123456789**

Univers LT Std
65 Bold

International style

苏黎世学派在德国的代表设计师是 Otl Aicher，以其参与创建的乌尔姆设计学院（Ulm School of Design）为教学地，他在艺术与设计之间划出了严格的界限，鼓励学生用数学的、逻辑的方法解决设计问题。Fig.81 是记录其一生中重要节点、作品、理念的书籍，其中，他为汉莎航空和 1972 年慕尼黑奥运会做的形象设计，是网格系统应用的典范。

从书中发现一个有意思的现象，初期，他的作品类似于威廉·莫里斯的自然人文风格——居中对称的衬线字体、手绘的图形。但 1947 年他来到瑞士，与国际风格的先驱 Max Bill 会面，同时也看到了 Müller-Brockmann 等人的作品。作品大多使用来自包豪斯的 Universal 与 Futura 字体，单色或两色，标准方形或双方形尺寸，构成感的几何图形，给了他很大的冲击。

之后他的风格大变，字体改用类似 Futura 的 Kristall Grotesque、Akzidenz Grotesk 等无衬线体，居中对称的排字方式变为散点分布。从中可以看出风格对设计师的影响，以及模仿风格是设计师未定型时的重要方法。

---

81. *Otl Aicher*，Markus Rathgeb 编，Phaidon Press，2015 年版。排版必读经典。

82. *Otl Aicher*，Markus Rathgeb 编，Phaidon Press，2015 年版，第 110–111 页。

81

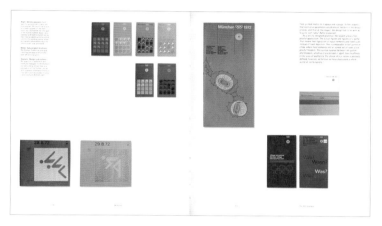

82

Wim Crouwel 是苏黎世学派在荷兰的代表设计师。最初他通过杂志 *TM*、*New Graphic Design* 中的文章了解国际风格，并在 1953 年去巴塞尔，见到了国际风格的代表人物 Armin Hofmann。但在 20 世纪 60 年代初期，国际风格在荷兰很受争议，被评论为"贫瘠、干如尘土、毫无生气""太严肃、太正式、毫无个性。虽然设计难免需要逻辑，但创造的新鲜感消失了"等。即便如此他还是拥抱国际风格，并成为继承者之一。

他的很多理念和作品在书籍 *Wim Crouwel-Modernist*（Fig.83）中都有介绍。比如，他认为设计师的任务是传达，信息必须以图形和非具象的方法视觉化。严格区分开艺术家与设计师，设计在于原因（Reason）和秩序（Order），是客观的、理性的，而非表达自身的感情。

Müller-Brockmann 的网格理念对他影响极大，他把单纯的文字，以严格的网格系统规划，通过大小、位置的安排，得到如图形般丰富的层次。他认为好设计是用简单直接的方式满足目的，创建一种 Structure 是其最重要的设计手法。

---

83. *Wim Crouwel-Modernist*，Frederike Huygen 著，Van Zoetendaal，2015 年版。排版必读经典。

84. *Wim Crouwel-Modernist*，Frederike Huygen 著，Van Zoetendaal，2015 年版，第 206–207 页。右页是他为 Stedelijk Museum 设计的作品，是其严格网格理念的代表作，很值得研究。

International style

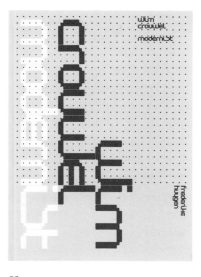

83

84

91

师从 Emil Ruder 的设计师 Helmut Schmid 是巴塞尔学派的继承者之一，他说人们普遍认为网格是瑞士风格的特点，但韵律（Rhythm）、秩序（Order）更重要。杉浦康平曾评价他："仔细聆听字母的声音，然后在白空间中准确地发现其应该出现的位置。"

因其在日本长时间的旅居经历，他的作品表现出一种带有东方柔和的理性，比如把日本枯山水的美学运用于排版中。他偏爱细体小写西文，很少用大写，喜欢用线条对信息进行分割，擅长从字间距的安排、词组的字号变化中得到韵律。

Fig.85 是他的著作 *Design is Attitude*，其中描述了一个通行的国际风格排版的练习方法。作者将其概括为三个阶段：限定（Restrict）、造形 (Structure)、选择（Select）。

首先限定开本、文本、字体、字号，保持这些元素固定，让注意力集中在下一步。然后单纯从形式上考虑文本长短、分段、位置、方向等的可能性，即便限制很多，这种可能性也是无限的。最后选择符合设计主题的可能性。一旦有了主题，也就有了主观意识。比如，要表达躁动还是安静，排版显然是不同的。

---

85. *Helmut Schmid: Design is Attitude*，Helmut Schmid 等著，Birkhauser，2007 年版。排版必读经典。

86. *Helmut Schmid: Design is Attitude*，Helmut Schmid 等著，Birkhauser，2007 年版，第 68-69 页。

helmut schmid:
gestaltung ist haltung
design is attitude

concept and design
fjodor gejko

birkhäuser

International style

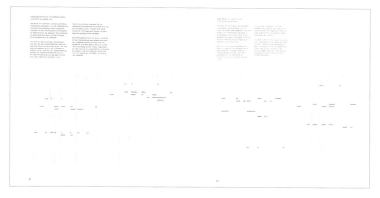

Emil Ruder 强调排印规则内的实验，受其影响的德国设计师 Wolfgang Weingart 则摆脱了排印规则的束缚，开创了瑞士朋克风（Swiss Punk Typography），著有 *Typography: My Way to Typography*，因其风格与典型的瑞士风格差别较大，这里不再赘述。

在这之后瑞士设计师 Willi Kunz 学习了 Wolfgang Weingart 在巴塞尔设计学院开设的课程，其风格介于瑞士风格与瑞士朋克风之间。Fig.87 是他著名的书籍 *Typography: Macro- and Microaesthetics*，其主张排版应有两层视角：宏观角度与微观角度。宏观的美感会引起读者最初的注意，然后引导其进入更复杂的微观细节上的美感。从他的作品中可以明显感受到对这种方法论的应用。

---

87. *Typography: Macro- and Microaesthetics*，Willi Kunz 著，Niggli，2002 年版。排版必读经典。

88. *Typography: Macro- and Microaesthetics*，Willi Kunz 著，Niggli，2002 年版，第 112–113 页。

International style

87

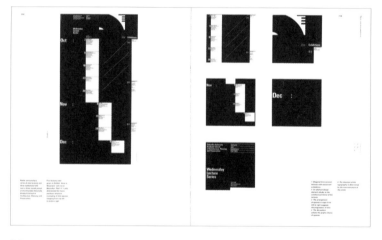

88

Signature or wordmark	The primary signature	The stacked wordmark
	The primary signature should be used whenever possible. The primary signature comprises the wordmark and the abbreviation.	The wordmark may be used alone in some situations. When the wordmark is used alone, the abbreviation may also appear somewhere in the publication or item.
	For the sake of consistency, please reproduce the art as it is supplied, and do not separate or alter the components.	It may be particularly useful to incorporate the wordmark separately from the abbreviation in display settings or in less formal situations.

The primary signature

The stacked wordmark

PROSPECT
AVENUE
CAPITAL

The abbreviation

1.1

VI 手册类似于使用说明，明晰、有条理地传达信息是第一要素，这很符合网格系统的特性。Fig.89 是作者为"PAC"设计的 VI 手册中的节选，纵向上用三列单元格，横向上单元格根据内容变化。

Signature or wordmark

One-line signature
The one-line signature should be used only in cases where the primary signature does not fit well, such as on the spine of a book.

The wordmark
The wordmark may be used without the abbreviation. These treatments are particularly useful for merchandise.

When the wordmark is used, the abbreviation should appear independently in the publication or on the item.

ONE-LINE SIGNATURE

P∧C PROSPECT ∧VENUE C∧PIT∧L

One-line wordmark

PROSPECT ∧VENUE C∧PIT∧L

1.2

Typographic style

タイポグラフィの
スタイル

排版的
风格

Constructional style
International style
Classical style
Grid system

左佐
编著

电子工业出版社

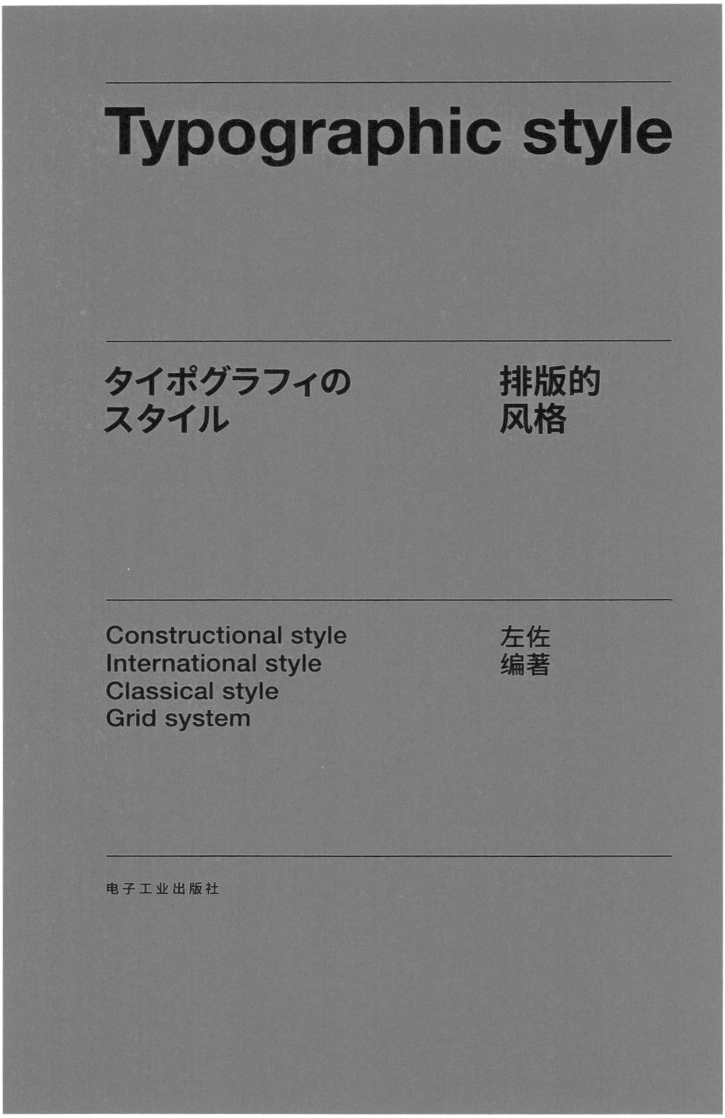

用国际风格设计的封面，更偏向于苏黎世学派，严格按照信息的传达顺序和网格来排版，即原因（Reason）与秩序（Order），无个人感情的表达。

Typographic style

タイポグラフィの
スタイル

排版的
风格

Constructional style
International style
Classical style
Grid system

左佐
编著

电子工业出版社

International style

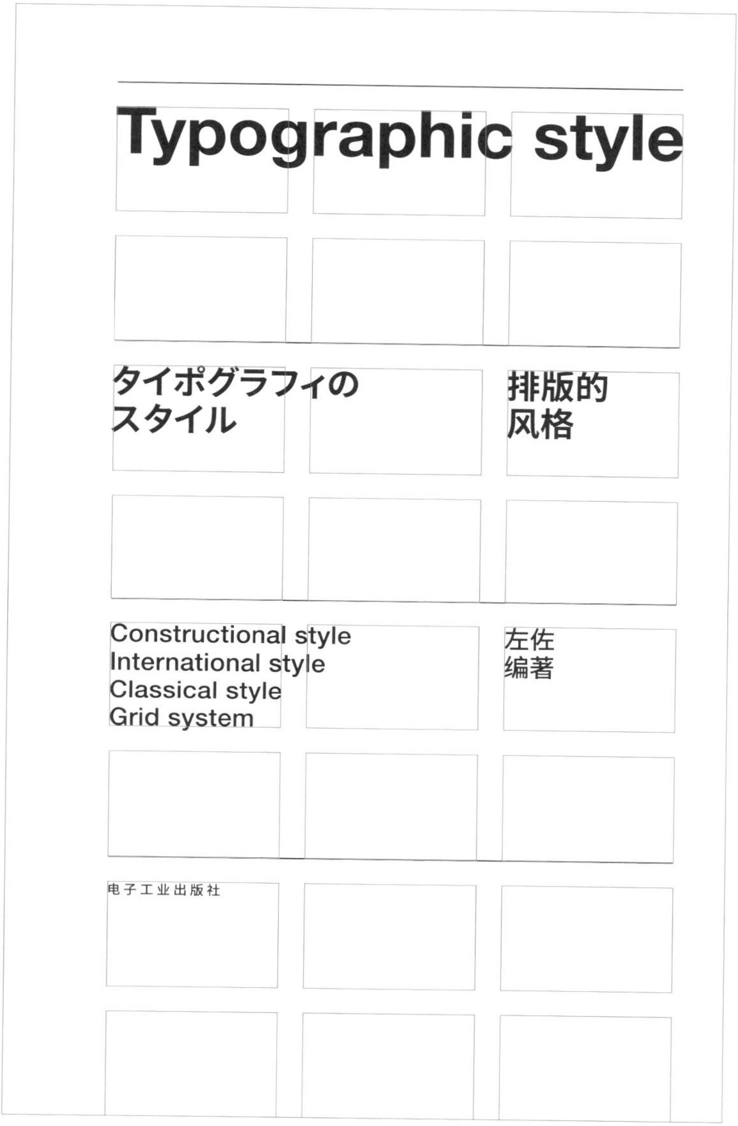

封面排版使用 3×8 的网格。西文字体选用 Helvetica，中文字体选用思源黑体。

typographic style

排版的风格

constructional style
international style
classical style
grid system

左佐编著

电子工业出版社

按照巴塞尔学派的方法，限定文本信息，简化模型。西文字体选用 Univers，字号相同，在空间中找到文本的位置，这种可能性是无限的。国际风格都讲求秩序（Order），但苏黎世学派更在乎原因（Reason），而巴塞尔学派更追求韵律（Rhyme）。

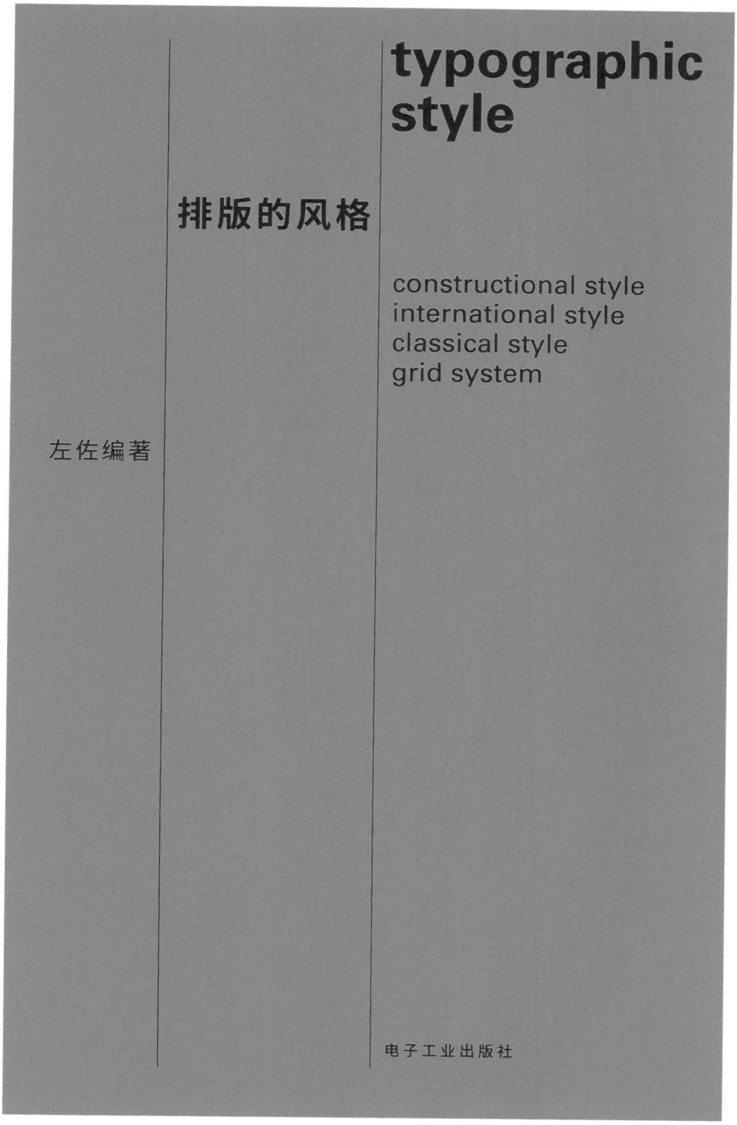

typographic style

排版的风格

constructional style
international style
classical style
grid system

左佐编著

电子工业出版社

对于文字位置的确定，除了要考虑主题的表达，更需要的是对文字本身品位的把握。把文字按阅读的先后顺序，用字重与竖线加以区分，进行封面排版。

流行风格

除了之前介绍的构成风格、国际风格、古典风格,每个时代都有特定的风格,但都衍生于三大基本风格,或是某个基本风格规则改变;或是某个基本风格在某个规则上走到极致;或是某个基本风格的规则向反方向走到极致;或是某两个或三个基本风格的融合。

叔本华说:欲望得不到满足就痛苦,欲望一旦满足就无聊,人生如同钟摆一样,在痛苦和无聊之间摇摆。这句话用来形容审美也很合适。审美在时间维度是一个轮回,某一个时间段人类追捧理性的国际风格,一旦这种风格遍地开花,随手可得,人类就会觉得无聊,会向着反方向探求。

当今流行的风格很受当下人类现状的影响。在看电影的选择上有一个普遍的观点,生活已经如此艰辛,为什么还要看那些压抑的、需要思考的电影,看些简单无脑的喜剧更轻松。在流行的风格上也类似,偏离正统的排版风格,是因为排版的规则实在太烦琐、太严肃,一切都从简、从轻、从乐,所以才催生了"新丑风""补丁风""冷对比"等流行风格。

逆风格时代呓语

所有的审美都在朝着相反的方向偏移，过去错误的现在是新奇的，过去不可以的现在是高级的，过去是丑陋的现在被供奉着。一场颠覆性的审美循环到了高潮。有人欢呼着、跳跃着，迎来属于自己的时代；有人试着欣赏、接受，意味着撕裂、迷失；有人大声地批判、大声地沉默，慢慢被"边缘"与"过时"淹没！！！

那些尖锐的、需要被避免的、需要修正的，统统以最为真实的形式赤裸裸地展示着它们不被习惯却被认为是独特的丑陋身体。在欢呼声中，一股浓重的、刺激的、直白的味道，告诉所有人习惯已经束缚了我们太久，我们需要错误、需要叛逆、需要一切与习惯不同的事物，我们要人为地颠覆、消灭、屠杀自己一手创立起来的"习惯"，向它们义无反顾地宣战！！！

那些留白的，终将被重新填满；那些精致的，都会被俗陋洗礼；那些过去以为是指引的星辰，全部被扭曲的风暴刮起的沙子遮盖；那些不愿看清潮流的车轮出现异动的人们，正被迅猛地、无声地、残忍地碾碎！！！

那些被我们抛弃的色彩，重新焕发出无法闪避的光芒；那些蹩脚的曲线，挣扎着占领一个又一个高地；那些被隐藏的缺陷，从腐败的土壤里爬起来，高声宣泄着压抑一万年的仇恨！！！

如果有一天基本的美学也被抛弃，我将死无葬身之地！！！

新丑风

当下流行的排版风格之一，是以日本的高田唯、中村至男、服部一成、仲条正义为代表，兴起的一股颠覆以往审美的排版风格，被称为 New Ugly（新丑风）。新丑风类似于构成风格，有一种原始粗犷感、刻意幼稚感，但人为精心安排的构成感较弱。

其特点很鲜明（Fig.94），中文字体偏爱早期为纸媒开发的、带喇叭口的黑体，或者一些很少人用的系统自带的宋体，英文偏爱 45 Light 的 Helvetica；排版规则趋近于幼稚模式，字间距为 0，行间距为字号大小，不分栏，文字瀑布式撑满版面；标题正文大部分无粗细大小区分，全部一个样式；矩形、方形、圆形、三角形等基本图形的色块作为装饰元素出现；箭头或者故意幼稚化的图形作为功能元素出现。

这种风格的产生有很大的时代因素。一方面可以从字体的新流行风格看出一点端倪。当下的审美趋势是一种极度简化后的轻盈传统风格。左图是传统字体 Palatino，右图是当下流行风格的字体 Noe Display，曲线造型的区别非常明显。表现在排版上也同样如此，那些传统条条框框的规则全部被极度简化。

graphic graphic

另一方面这可能是对高品位设计过度普及的逆反。当今信息传播非常便利，一种风格被认可后，很快会被很多人效仿，进而出现设计大范围趋同的情况。熟能生巧，亦能生厌，当高品位因过于频繁出现而导致视觉疲劳时，设计师掉头迈向另一个极端，在广大的市井中寻求风格。比如，Fig.95 是街头很常见的小广告，但只需稍加润色就是一幅大俗大雅的流行风格作品。

Popular style

94

95

补丁风

新丑风加入颜色、构成、拼贴方式的变异，也可以说是街头小广告的演化，以被矩形裁切的大字放置于矩形色块之上，或以小字置于矩形色块之内；颜色偏爱饱和度不高的脏色，或者饱和度最高的 CMYK 印刷基本色（Fig.96）。矩形色块之间也可以产生层叠关系（Fig.99），文字越大被裁切的信息越多，被裁切的信息过多，会重复出现被裁切信息的矩形，但信息文字变小，被裁切的部分也会变少，直至可完整阅读信息。

冷对比

还有一种流行于西方的新排版方式，去掉了新丑风中的刻意幼稚化、丑化的形式，以极度简化的排版方式出现。一般信息被分为两层（Fig.97），一层是呈瀑布式排列的文字信息，一层是很大的图形或图片。一种本应是点的文字，因紧凑密布形成面的形式；而本应是面的图形图像，因尺寸巨大，细节上呈现出点、线的形式。这两种反差对比，构成了这种风格的独特美感。

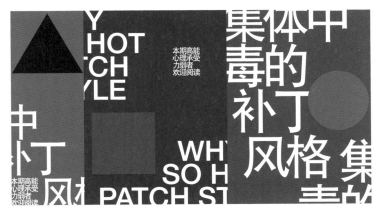

用流行风格设计的封面。新丑风中紧密的行距、补丁风中的色块、冷对比中偏细的字形，但又结合基本的构成风格、国际风格，使版式避免过于主观化，向原因与秩序靠拢。封面的网格由 Jan Tschichold 的作品（Fig.40）演化而来。

用流行风格设计的封面。文字被层叠的矩形剪切得支离破碎,
但仔细看仍能拼凑出完整的信息,颜色使用饱和度不高的脏色。

西文网格

什么是网格？简单来说，把版心划分为若干单元格，就是网格。网格包含标题、正文、注释、图注、图片的复合信息，可以用网格梳理出客观、清晰、具有阅读逻辑、具有韵律的排版形式。

单位换算，现在常用的是，1pt=1/72 in≈0.35mm，作者的习惯是在确定了开本之后，在软件"首选项—单位"中把单位设置为"点（pt）"，之后在网格设定中会非常方便，比如印刷出血 3mm，与 10pt（3.5mm）相当。先以西文为例解说网格。

<u>古典风格的网格</u>
古典风格的网格一般正文不分栏，且各类信息有比较固定的位置和大小关系，相对来说比较容易构建。所以我们先以此开始，通过原尺寸的内页样例，一步一步解析构建的过程。

1. 开本尺寸选择 270pt×431pt（95mm×152mm），长宽比接近黄金比例，上图为原始尺寸。按古典风格奥义设置版心。

Typography is the art and technique of arranging type to make written language legible, readable, and appealing when displayed. The arrangement of type involves selecting typefaces, point sizes, line lengths, line-spacing (leading), and letter-spacing (tracking), and adjusting the space between pairs of letters (kerning[1]).

The term typography is also applied to the style, arrangement, and appearance of the letters, numbers, and symbols created by the process. Type design is a closely related craft, sometimes considered part of typography; most typographers do not design typefaces, and some type designers do not consider themselves typographers.[2] Typography also may be used as a decorative device, unrelated to communication of information.

Typography is the work of typesetters (also known as compositors), typographers, graphic designers, art directors, manga artists, comic book artists, graffiti artists, and, now, anyone who arranges words, letters, numbers, and symbols for publication, display, or distribution, from clerical workers and newsletter writers to self-publishing materials.

Digitization opened up typography to new generations of previously unrelated designers and lay users. As the capability to create typography has become ubiquitous, the application of principles and best practices developed over generations of skilled workers and professionals has diminished. So at a time when scientific techniques can support the proven traditions through understanding the limitations of human vision, typography as often encountered may fail to achieve its principal objective: effective communication.

2. 设定正文字号、行距。这里使用字体 Garamond，字号为 8pt，行距为 10pt，放入版心。正文为 29 行时，会溢出版心一点（一般都不会正好容下整数行）。

Western grid system

Typography is the art and technique of arranging type to make written language legible, readable, and appealing when displayed. The arrangement of type involves selecting typefaces, point sizes, line lengths, line-spacing (leading), and letter-spacing (tracking), and adjusting the space between pairs of letters (kerning[1]).

The term typography is also applied to the style, arrangement, and appearance of the letters, numbers, and symbols created by the process. Type design is a closely related craft, sometimes considered part of typography; most typographers do not design typefaces, and some type designers do not consider themselves typographers.[2] Typography also may be used as a decorative device, unrelated to communication of information.

Typography is the work of typesetters (also known as compositors), typographers, graphic designers, art directors, manga artists, comic book artists, graffiti artists, and, now, anyone who arranges words, letters, numbers, and symbols for publication, display, or distribution, from clerical workers and newsletter writers to self-publishing materials.

Digitization opened up typography to new generations of previously unrelated designers and lay users. As the capability to create typography has become ubiquitous, the application of principles and best practices developed over generations of skilled workers and professionals has diminished. So at a time when scientific techniques can support the proven traditions through understanding the limitations of human vision, typography as often encountered may fail to achieve its principal objective: effective communication.

3. 如果以易读性为先，则需要调整版心高度，使其正好包含 29 行。如果以形式感为先，则可调整字号与行距，使整数行适应版心高度。

4. 计算调整后版心高度: (29−1)×10pt+8pt=288pt, 即（行数−1)×行距+字号＝版心高度。

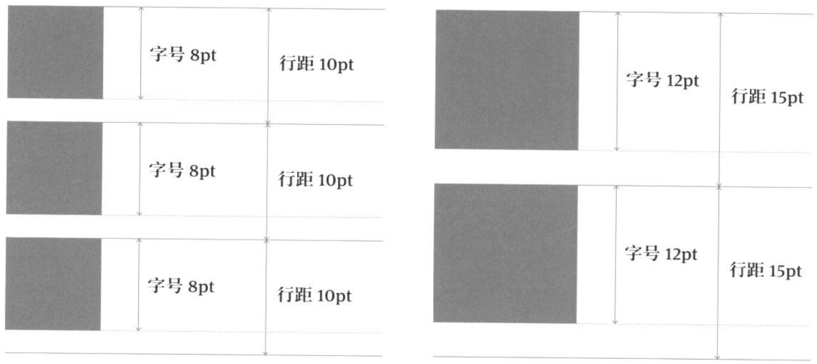

Western grid system

100

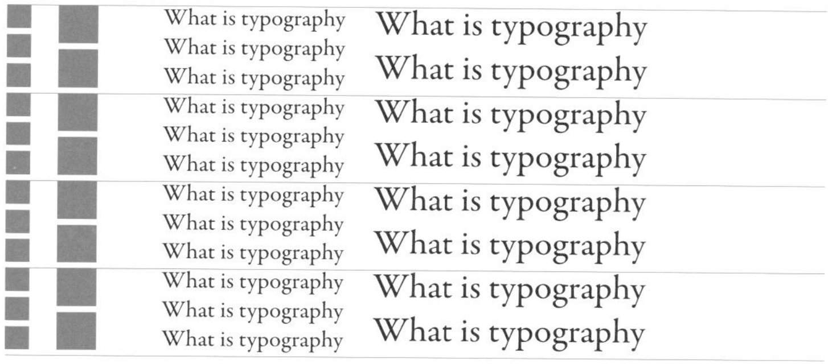

101

5. 版心和正文确定后，根据正文的字号和行距，设定标题、注解的字号和行距。为了让标题、注解与正文间形成韵律美，会用到简单的数学计算。

Fig.100 中三行正文高度：3×10pt=30pt，与两行标题高度相同，可以得出标题的行距：30pt/2=15pt，那么标题字号可以设定在 8pt 与 15pt 之间。这里设定标题字号为 12pt，行距为 15pt。Fig.101 是标题与正文字号的视觉韵律。

102

103

6. 更大的标题设定。Fig.102 中两行正文高度：2×10pt=20pt，与一行标题高度相同，可以得出标题的行距为 20pt，那么标题字号可以设定在 8pt 与 20pt 之间。这里设定标题字号为 16pt，行距为 20pt。Fig.103 是字号之间的韵律。

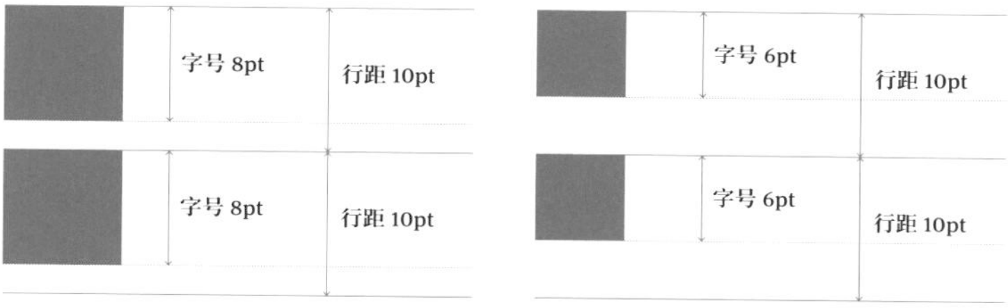

104

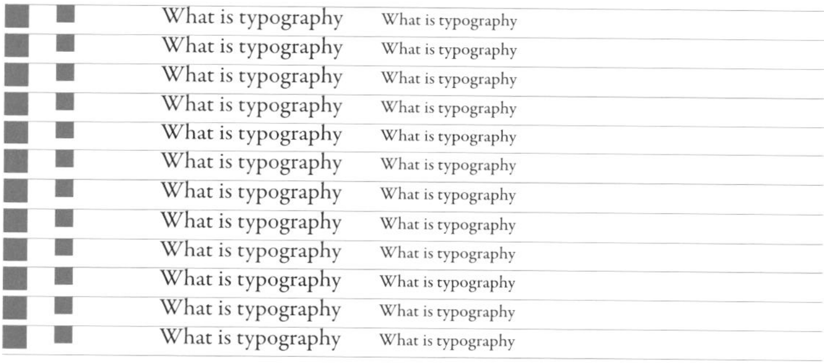

105

7. 以同样的方式设定注释字号（Fig.104），古典风格注释字号比正文小 2pt，行距相同，即字号为 6pt，行距为 10pt。Fig.105 是字号之间的韵律。

What is typography

TYPOGRAPHY is the art and technique of arranging type to make written language legible, readable, and appealing when displayed. The arrangement of type involves selecting typefaces, point sizes, line lengths, line-spacing (leading), and letter-spacing (tracking), and adjusting the space between pairs of letters (kerning[1]).

Arrangement of letters

The term typography is also applied to the style, arrangement, and appearance of the letters, numbers, and symbols created by the process. Type design is a closely related craft, sometimes considered part of typography; most typographers do not design typefaces, and some type designers do not consider themselves typographers.[2] Typography also may be used as a decorative device, unrelated to communication of information.

Typography is the work of typesetters (also known as compositors), typographers, graphic designers, art directors, manga artists, comic book artists, graffiti artists, and, now, anyone who arranges words, letters, numbers, and symbols for publication, display, or distribution, from clerical workers and newsletter writers to self-publishing materials.

1. Bringhurst, Robert. *The Elements of Typographic Style*. Hartley & Marks.
2. Pipes, Alan. *Production For Graphic Designers*. Prentice-Hall.

15

Fig.106 是用古典风格的规则排版的页面，字体用 Garamond，标题字号为 12pt，章节首行第一个单词用大型小写，正文字号为 8pt，段首缩进 8pt，副标题字号为 8pt 的 Italic 体，注解字号为 6pt，行首缩进 8pt。

WHAT IS TYPOGRAPHY

Typography is the art and technique of arranging type to make written language legible, readable, and appealing when displayed. The arrangement of type involves selecting typefaces, point sizes, line lengths, line-spacing (leading), and letter-spacing (tracking), and adjusting the space between pairs of letters (kerning).

ARRANGEMENT OF LETTERS

symbols

The term typography is also applied to the style, arrangement, and appearance of the letters, numbers, and symbols created by the process. Type design is a closely related craft, sometimes considered part of typography; most typographers do not design typefaces, and some type designers do not consider themselves typographers. Typography also may be used as a decorative device, unrelated to communication of information.

composition

Typography is the work of typesetters (also known as compositors), typographers, graphic designers, art directors, manga artists, comic book artists, graffiti artists, and, now, anyone who arranges words, letters, numbers, and symbols

15

Fig.107 是另一种用古典风格的规则排版的页面，字体用 Garamond，大标题用大写字母，字号为 10pt，次标题用大型小写字母，字号为 8pt，小标题用 Italic 体，字号为 8pt，正文字号为 8pt，段首缩进 8pt。

WHAT IS TYPOGRAPHY

Typography is the art and technique of arranging type to make written language legible, readable, and appealing when displayed. The arrangement of type involves selecting typefaces, point sizes, line lengths, line-spacing (leading), and letter-spacing (tracking), and adjusting the space between pairs of letters (kerning[1]).

Arrangement

The term typography is also applied to the style, arrangement, and appearance of the letters, numbers, and symbols created by the process. Type design is a closely related craft, sometimes considered part of typography; most typographers do not design typefaces, and some type designers do not consider themselves typographers.[2] Typography also may be used as a decorative device, unrelated to communication of information.

Composition

Typography is the work of typesetters (also known as compositors), typographers, graphic designers, art directors, manga artists, comic book artists, graffiti artists, and, now, anyone who arranges words, letters, numbers, and symbols for publication, display, or distribution, from clerical workers and newsletter writers to self-publishing materials.

Designers

Digitization opened up typography to new generations of previously unrelated designers and lay users. As the capability to create typography has become ubiquitous.

1. Bringhurst Robert. *The Elements of Typographic Style*, Hartley & Marks
2. Pipes Alan, *Production For Graphic Designers*, Prentice-Hall

15

Fig.108 是无章节标题时的页面，字体用 Garamond，左页天头写书籍名称，字号为 8pt，大型小写；右页天头写章节名称，字号为 8pt，大型小写。外边距可用 6pt 的 Italic 体标注长段落的概要。

Composition

Typography is the work of typesetters (also known as compositors), typographers, graphic designers, art directors, manga artists, comic book artists, graffiti artists, and, now, anyone who arranges words, letters, numbers, and symbols for publication, display, or distribution, from clerical workers and newsletter writers to self-publishing materials.

Principles

Digitization opened up typography to new generations of previously unrelated designers and lay users. As the capability to create typography has become ubiquitous, the application of principles and best practices developed over generations of skilled workers and professionals has diminished. So at a time when scientific techniques can support the proven traditions through understanding the limitations of human vision, typography as often encountered may fail to achieve its principal objective: effective communication.

Typography is the art and technique of arranging type to make written language legible, readable, and appealing when displayed. The arrangement of type involves selecting typefaces, point sizes, line lengths, line-spacing (leading), and letter-spacing (tracking), and adjusting the space between pairs of letters.

Numbers

The term typography is also applied to the style, arrangement, and appearance of the letters, numbers, and symbols created by the process. Type design is a closely related craft, sometimes considered part of typography; most typographers do not design typefaces, and some type designers do not consider themselves typographers. Typography also may be used as a decorative device, unrelated to communication of information.

16

WHAT IS TYPOGRAPHY

Typography is the art and technique of arranging type to make written language legible, readable, and appealing when displayed. The arrangement of type involves selecting typefaces, point sizes, line lengths, line-spacing (leading), and letter-spacing (tracking), and adjusting the space between pairs of letters (kerning[1]).

Arrangement

The term typography is also applied to the style, arrangement, and appearance of the letters, numbers, and symbols created by the process. Type design is a closely related craft, sometimes considered part of typography; most typographers do not design typefaces, and some type designers do not consider themselves typographers.[2] Typography also may be used as a decorative device, unrelated to communication of information.

Composition

Typography is the work of typesetters (also known as compositors), typographers, graphic designers, art directors, manga artists, comic book artists, graffiti artists, and, now, anyone who arranges words, letters, numbers, and symbols for publication, display, or distribution, from clerical workers and newsletter writers to self-publishing materials.

Designers

Digitization opened up typography to new generations of previously unrelated designers and lay users. As the capability to create typography has become ubiquitous.

1. Bringhurst Robert. *The Elements of Typographic Style*, Hartley & Marks
2. Pipes Alan, *Production For Graphic Designers* , Prentice-Hall

古典风格的网格线。

COMPOSITION

Typography is the work of typesetters (also known as compositors), typographers, graphic designers, art directors, manga artists, comic book artists, graffiti artists, and, now, anyone who arranges words, letters, numbers, and symbols for publication, display, or distribution, from clerical workers and newsletter writers to self-publishing materials.

Principles

Digitization opened up typography to new generations of previously unrelated designers and lay users. As the capability to create typography has become ubiquitous, the application of principles and best practices developed over generations of skilled workers and professionals has diminished. So at a time when scientific techniques can support the proven traditions through understanding the limitations of human vision, typography as often encountered may fail to achieve its principal objective: effective communication.

Typography is the art and technique of arranging type to make written language legible, readable, and appealing when displayed. The arrangement of type involves selecting typefaces, point sizes, line lengths, line-spacing (leading), and letter-spacing (tracking), and adjusting the space between pairs of letters.

Numbers

The term typography is also applied to the style, arrangement, and appearance of the letters, numbers, and symbols created by the process. Type design is a closely related craft, sometimes considered part of typography; most typographers do not design typefaces, and some type designers do not consider themselves typographers. Typography also may be used as a decorative device, unrelated to communication of information.

国际风格的网格

国际风格的网格,特点是根据内容会分多栏,内容呈不对称散点排布,开本以正方形居多。下面详细介绍网格的设置过程。

---

1. 设定开本。根据正六边形得出一个国际风格偏爱的方形。开本尺寸:270pt×312pt(95mm×110mm),Fig.110 为原始尺寸。

2. 确定正文字体、字号和行距。这里字体使用 Univers,字号为 6pt,行距为 7.5pt。然后置入页面(Fig.110),超过 41 行时正文溢出,所以页面可容纳 41 行。

3. 设定版心高度。在设定前,先了解行数与版心高度的数学关系。Fig.111 的版心高度被分为 3 个单元格,每个单元格容纳 4 行,单元格间隔 1 行,那么版心可容纳:3×4+(3−1)=14 行,版心高度也就有了,行数乘以行距即可。所以版心高度与行数的关系为:版心行数 = 单元格数 × 单元格行数 +(单元格数−1)。

Western grid system

```
1
2
3
4
5
6
7
8
9
10
11
12
13
14
15
16
17
18
19
20
21
22
23
24
25
26
27
28
29
30
31
32
33
34
35
36
37
38
39
40
41
42
```

110

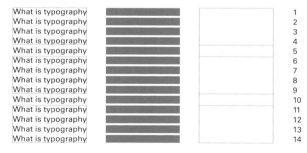

111

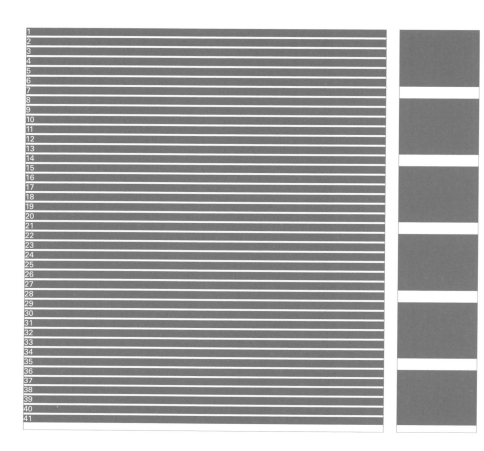

4. 设定版心高度。这里纵向选择划分为 6 个单元格，通过步骤 2 知道页面最多可容纳 41 行，通过步骤 3 的公式：6× 单元格行数 +(6−1)=41，得出单元格行数为 6，如果含小数点，只取整数。也就是说当页面纵向划分为 6 个单元格时，每个单元格最多可容纳 6 行正文。

5. 设定版心高度。从步骤 4 得知，当纵向划分为 6 个单元格时，每个单元格至多容纳 6 行正文，多于 6 行时会溢出页面，少于 6 行时，上下留白会增大。所以可在 1~6 之间调整单元格可容纳的行数，得出留白适当的版面。上图是单元格行数为 5 时，得到的版心高度。

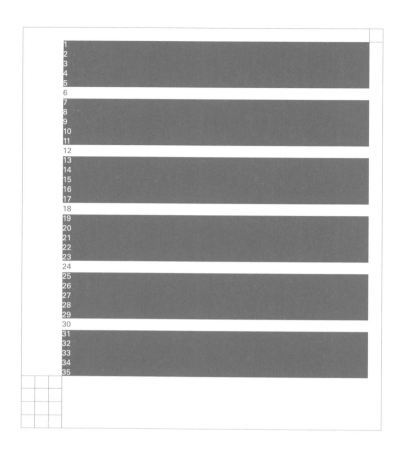

6. 设定页边距。为了让上下左右边距形成数学比例关系，用页面高度减去版心高度，然后将其 5 等分（见上页图），数值可根据情况设定，取 1 份作为上边距，4 份作为下边距；左边距取 3 份，右边距取 1 份。

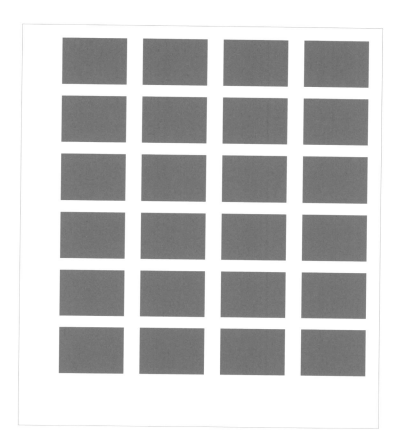

7. 设定横向单元格。这里横向取 4 个单元格，横向单元格的间距不必与纵向单元格间距相同，一般至少为行距的大小。这里取字号的 2 倍，12pt。

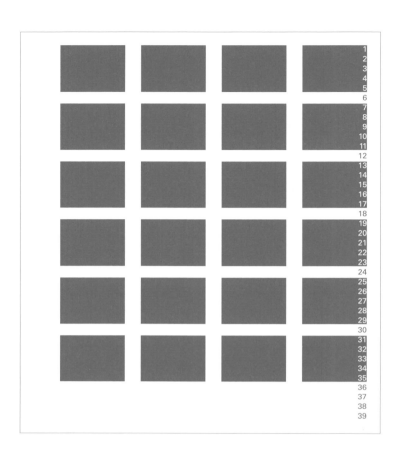

8. 确定页码位置。这里设定在右下角，正文延伸后的第 39 行的位置。

Western grid system

9. 至此一个页面的基础网格就建立完成。网格的每一部分
都来自最初的开本、正文，富有理性、数学逻辑和韵律感。

What is Typography
Art and technique

Typography is the art and technique of arranging type to make written language legible, readable, and appealing when displayed. The arrangement of type involves selecting typefaces, point sizes, line lengths, line-spacing (leading), and letter-spacing (tracking), and adjusting the space between pairs of letters (kerning).

The term typography is also applied to the style, arrangement, and appearance of the letters, numbers, and symbols created by the process. Type design is a closely related craft, sometimes considered part of typography; most typographers do not design typefaces, and some type designers do not consider themselves typographers. Typography also may be used as a decorative device, unrelated to communication of information.

Typography is the work of typesetters (also known as compositors), typographers, graphic designers, art directors, manga artists, comic book artists, graffiti artists, and, now, anyone who arranges words, letters, numbers, and symbols for publication, display, or distribution, from clerical workers and newsletter writers to self-publishing materials.

Digitization opened up typography to new generations of previously unrelated designers and lay users. As the capability to create typography has become ubiquitous, the application of principles and best practices developed over generations of skilled workers and professionals has diminished. So at a time when scientific techniques can support the proven traditions through understanding the limitations of human vision, typography as often encountered may fail to achieve its principal objective: effective communication.

15

利用古典风格根据正文确定标题的方法，确定标题字号为9pt，行距为11.25pt；排版方式按苏黎世学派风格，标题用 Univers 65 Bold，正文左右强制对齐，段落间不空行，以达到形式上的构成感。

What is Typography

Art and technique

Typography is the art and technique of arranging type to make written language legible, readable, and appealing when displayed. The arrangement of type involves selecting typefaces, point sizes, line lengths, line-spacing (leading), and letter-spacing (tracking), and adjusting the space between pairs of letters (kerning).

The term typography is also applied to the style, arrangement, and appearance of the letters, numbers, and symbols created by the process. Type design is a closely related craft, sometimes considered part of typography; most typographers do not design typefaces, and some type designers do not consider themselves typographers. Typography also may be used as a decorative device, unrelated to communication of information.

Typography is the work of typesetters (also known as compositors), typographers, graphic designers, art directors, manga artists, comic book artists, graffiti artists, and, now, anyone who arranges words, letters, numbers, and symbols for publication, display, or distribution, from clerical workers and newsletter writers to self-publishing materials.

Digitization opened up typography to new generations of previously unrelated designers and lay users. As the capability to create typography has become ubiquitous, the application of principles and best practices developed over generations of skilled workers and professionals has diminished. So at a time when scientific techniques can support the proven traditions through understanding the limitations of human vision, typography as often encountered may fail to achieve its principal objective: effective communication.

15

隐藏网格后的排版形式。

		Typography is the art and technique of arranging type to make written language legible, readable, and appealing when displayed. The arrangement of type involves selecting typefaces, point sizes, line lengths, line-spacing (leading), and letter-spacing (tracking), and adjusting the space between pairs of letters (kerning).
What is Typography		The term typography is also applied to the style, arrangement, and appearance of the letters, numbers, and symbols created by the process. Type design is a closely related craft, sometimes considered part of typography; most typographers do not design typefaces, and some type designers do not consider themselves typographers. Typography also may be used as a decorative device, unrelated to communication of information.
Typography is the art and technique of arranging type		
		Typography is the work of typesetters (also known as compositors), typographers, graphic designers, art directors, manga artists, comic book artists, graffiti artists, and, now, anyone who arranges words, letters, numbers, and symbols for publication, display, or distribution, from clerical workers and newsletter writers to self-publishing materials.
		Digitization opened up typography to new generations of previously unrelated designers and lay users. As the capability to create typography has become ubiquitous, the application of principles and best practices developed over generations of skilled workers and professionals has diminished. So at a time when scientific techniques can support the proven traditions through understanding the limitations of human vision, typography as often encoun-

15

网格设定为 3 列单元格。排版方式按巴塞尔学派风格，标题字号为 9pt，行距为 11.25pt，字体与正文相同（Univers 55 Roman）；正文左对齐，右侧参差分布；段落间空行；页码移至左下角，字体同样与正文相同。强化不对称的排版形式。

134

What is Typography

Typography is the art and technique of arranging type

Typography is the art and technique of arranging type to make written language legible, readable, and appealing when displayed. The arrangement of type involves selecting typefaces, point sizes, line lengths, line-spacing (leading), and letter-spacing (tracking), and adjusting the space between pairs of letters (kerning).

The term typography is also applied to the style, arrangement, and appearance of the letters, numbers, and symbols created by the process. Type design is a closely related craft, sometimes considered part of typography; most typographers do not design typefaces, and some type designers do not consider themselves typographers. Typography also may be used as a decorative device, unrelated to communication of information.

Typography is the work of typesetters (also known as compositors), typographers, graphic designers, art directors, manga artists, comic book artists, graffiti artists, and, now, anyone who arranges words, letters, numbers, and symbols for publication, display, or distribution, from clerical workers and newsletter writers to self-publishing materials.

Digitization opened up typography to new generations of previously unrelated designers and lay users. As the capability to create typography has become ubiquitous, the application of principles and best practices developed over generations of skilled workers and professionals has diminished. So at a time when scientific techniques can support the proven traditions through understanding the limitations of human vision, typography as often encoun-

15

Western grid system

隐藏网格后的排版形式。

Typography is the art and technique of arranging type to make written language legible, readable, and appealing when displayed. The arrangement of type involves selecting typefaces, point sizes, line lengths, line-spacing (leading), and letter-spacing (tracking), and adjusting the space between pairs of letters (kerning).

What is Typography

The term typography is also applied to the style, arrangement, and appearance of the letters, numbers, and

15

按巴塞尔学派的实验理念，在网格内得出更多的形式变化。注意这时的实验大多以个人品位为主，可脱离开内容本身的传达。标题字号放大至 20pt，行距为 22.5pt，增大对比，增大留白。巴塞尔学派不太受网格束缚，标题感觉偏上，可以下移一行。

Western grid system

Typography is the art and technique of arranging type to make written language legible, readable, and appealing when displayed. The arrangement of type involves selecting typefaces, point sizes, line lengths, line-spacing (leading), and letter-spacing (tracking), and adjusting the space between pairs of letters (kerning).

The term typography is also applied to the style, arrangement, and appearance of the letters, numbers, and

What is Typography

15

隐藏网格后的排版形式。可见标题并未像左页一样依据网格放置，而是根据个人感觉向下移动。这也是巴塞尔学派更注重个人感觉、品位的特点，是理性的理论不可实现的，只能通过不断地训练提升个人品位来达成。下移的原因也可用构成风格的方形原则来解释，下移后中心的留白空间更接近正方形。

What is Typography	Typography is the art and technique of arranging type to make written language legible, readable, and appealing when displayed. The arrangement of type involves selecting typefaces, point sizes, line lengths, line-spacing (leading), and letter-spacing (tracking), and adjusting the space between pairs of letters (kerning).
Typography is the art and technique of arranging type	The term typography is also applied to the style, arrangement, and appearance of the letters, numbers, and symbols created by the process. Type design is a closely related craft, sometimes considered part of typography; most typographers do not design typefaces, and some type designers do not consider themselves typographers. Typography also may be used as a decorative device, unrelated to communication of information.
	Typography is the work of typesetters (also known as compositors), typographers, graphic designers, art directors, manga artists, comic book artists, graffiti artists, and, now, anyone who arranges words, letters, numbers, and symbols for publication, display, or distribution, from clerical workers and newsletter writers to self-publishing materials.
	Digitization opened up typography to new generations of previously unrelated designers and lay users. As the capability to create typography has become ubiquitous, the application of principles and best practices developed over generations of skilled workers and professionals has diminished. So at a time when scientific techniques can support the proven traditions through understanding the limitations of human vision, typography as often encoun-

15

组合起来的排版形式。从头回想，会发现每一个细节都与整体有着隐藏的数学逻辑关系，而这些数学关系又是基于内容表达、视觉形式得出的。规则之下的感觉判定，感觉之上的规则建立。

Western grid system

What is Typography

Typography is the art and technique of arranging type to make written language legible, readable, and appealing when displayed. The arrangement of type involves selecting typefaces, point sizes, line lengths, line-spacing (leading), and letter-spacing (tracking), and adjusting the space between pairs of letters (kerning).

The term typography is also applied to the style, arrangement, and appearance of the letters, numbers, and

中文网格

中文的行宽

每个汉字的宽度、高度,与字号大小相等,所以其组成的行的宽度,不像西文单词组成的行随机不可计算。比如下图,字号为 34pt,每行 10 个字,行宽是 10×34pt=340pt。

中文字号的基数

比如基数为 2pt,正文字号可设定为 4×2pt=8pt,标题可设定为 6×2pt=12pt,注解可设定为 3×2pt=6pt。这样汉字会在列之间形成数学韵律(见下图),而不仅是行之间。比如基数设定为 1.5pt,则得到 4×1.5pt=6pt,5×1.5pt=7.5pt,6×1.5pt=9pt,7×1.5pt=10.5pt……可从中选择适合文本的字号使用。

<u>中文的行距</u>

西文字母有上伸部分、下伸部分，而汉字则简单很多，上下部分都是顶格排列。汉字行距常用倍数来表示，比如下图，字号为 12pt 的文本，分别设定 1 倍（12pt）、1.5 倍 (18pt)、2 倍 (24pt) 的行距。

字字字字字字字字字字字字字字字字
字字字字字字字字字字字字字字字字
字字字字字字字字字字字字字字字字

字字字字字字字字字字字字字字字字
字字字字字字字字字字字字字字字字
字字字字字字字字字字字字字字字字

字字字字字字字字字字字字字字字字
字字字字字字字字字字字字字字字字
字字字字字字字字字字字字字字字字

可以发现当行距为 2 倍时，两行之间的空隙与行的高度相同，也就是说版心可以划分为字号大小的单元格。虽然这种行距会稍显稀疏，但可以在很大程度上简化汉字网格的设定（2 倍行距的网格可参阅刘晓翔先生的《汉字排版》）。

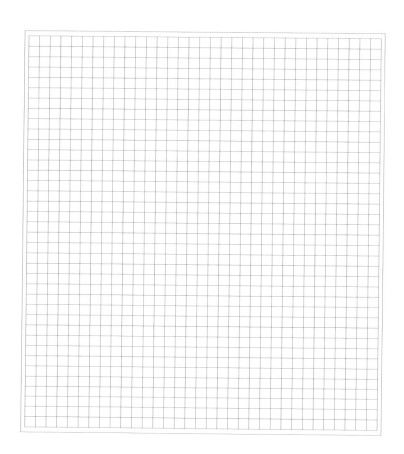

<u>2 倍行距的中文网格</u>

1. 用国际风格网格中设定的开本，字号基数设定为 2pt，正文为 4×2pt=8pt，标题为 6×2pt=12pt，注解为 3×2pt=6pt，行距为 2 倍。首先用正文单个汉字大小的 8pt×8pt 方格填满页面。

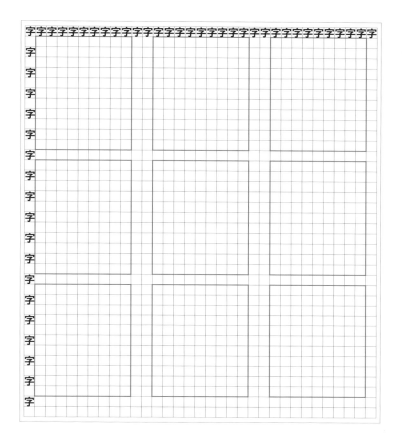

2. 确定版心。因整个页面被分为大小相同的文字格，版心和网格的确定就会很简单。比如想要一个 3×3 的网格。

| 什么是西文排印 | 西文排印是一种通过安排字体，使文本的展示具有可读性、易读性、诱目性的艺术和技术。这种安排包含选择字体，设定字号、行长、行间距，以及字母间距。术语西文排印也会用于在此过程中产生的字母、数字、符号的风格、安排、外观。字体设计与西文排印关系密切，有时也会作为西文排印的一部分。大部分西文排印师并不设计字体，而一些字体设计师并不认为自己是西文排印师。西文排印也会被用来指与信息传达无关的装饰图案。|
|What is Typography| 任何涉及到安排单词、字母、数字和符号以出版、展示的工作，都在西文排印的范围内，比如组版师、排印师、平面设计师、艺术从业者、动漫师、涂鸦师，甚至公司文员、报刊作者等。电子时代之前，西文排印一直是一个专业领域。数字化把西文排印扩大到与设计无关的新人群。随着西文排印的能力变得越来越普遍，|
| 15 | 由专业西文排印者代代相传的规则应用与训 |

3. 简化模型，中文标题、正文全部用相同字号、相同行距，置入文本之后的排版形式。西文部分，字号为 8pt，3 个行距的西文与 2 个行距的中文等高，西文行距为 2×8pt/3=10.7pt。

| 什么是西文排印 | 西文排印是一种通过安排字体，使文本的展示具有可读性、易读性、诱目性的艺术和技术。这种安排包含选择字体，设定字号、行长、行间距，以及字母间距。术语西文排印也会用于在此过程中产生的字母、数字、符号的风格、安排、外观。字体设计与西文排印关系密切，有时也会作为西文排印的一部分。大部分西文排印师并不设计字体，而一些字体设计师并不认为自己是西文排印师。西文排印也会被用来指与信息传达无关的装饰图案。 |
| What is Typography | 任何涉及到安排单词、字母、数字和符号以出版、展示的工作，都在西文排印的范围内，比如组版师、排印师、平面设计师、艺术从业者、动漫师、涂鸦师，甚至公司文员、报刊作者等。电子时代之前，西文排印一直是一个专业领域。数字化把西文排印扩大到与设计无关的新人群。随着西文排印的能力变得越来越普遍，由专业西文排印者代代相传的规则应用与训 |

隐藏网格后的排版形式。

什么是
西文排印

What is
Typography

西文排印是一种通过安排字体，使文本的展示具有可读性、易读性、诱目性的艺术和技术。

15

西文排印是一种通过安排字体，使文本的展示具有可读性、易读性、诱目性的艺术和技术。这种安排包含选择字体，设定字号、行长、行间距，以及字母间距。术语西文排印也会用于在此过程中产生的字母、数字、符号的风格、安排、外观。字体设计与西文排印关系密切，有时也会作为西文排印的一部分。大部分西文排印师并不设计字体，而一些字体设计师并不认为自己是西文排印师。西文排印也会被用来指与信息传达无关的装饰图案。

任何涉及到安排单词、字母、数字和符号以出版、展示的工作，都在西文排印的范围内，比如组版师、排印师、平面设计师、艺术从业者、动漫师、涂鸦师，甚至公司文员、报刊作者等。电子时代之前，西文排印一直是一个专业领域。数字化把西文排印扩大到与设计无关的新人群。随着西文排印的能力变得越来越普遍，由专业西文排印者代代相传的规则应用与训

4. 标题、注解加入变化。中文标题字号为 12pt，行距为 24pt。西文标题的行距设定为 16pt，字号设定为 12pt。注解字号为 6pt，行距为 12pt。

什么是
西文排印

What is
Typography

西文排印是一种通过安排字体，使文本的展示具有可读性、易读性、诱目性的艺术和技术。

西文排印是一种通过安排字体，使文本的展示具有可读性、易读性、诱目性的艺术和技术。这种安排包含选择字体，设定字号、行长、行间距，以及字母间距。术语西文排印也会用于在此过程中产生的字母、数字、符号的风格、安排、外观。字体设计与西文排印关系密切，有时也会作为西文排印的一部分。大部分西文排印师并不设计字体，而一些字体设计师并不认为自己是西文排印师。西文排印也会被用来指与信息传达无关的装饰图案。

任何涉及到安排单词、字母、数字和符号以出版、展示的工作，都在西文排印的范围内，比如组版师、排印师、平面设计师、艺术从业者、动漫师、涂鸦师，甚至公司文员、报刊作者等。电子时代之前，西文排印一直是一个专业领域。数字化把西文排印扩大到与设计无关的新人群。随着西文排印的能力变得越来越普遍，由专业西文排印者代代相传的规则应用与训

隐藏网格后的排版形式。这时会发现随着字号的增大，2倍行距的稀疏感被放大。中文标题行距明显过大，可以从功能性出发，调整至视觉舒服的行距；也可以从形式感出发，不做调整，保持数学上的绝对韵律关系。

| 什么是
西文排印 | 西文排印是一种通过安排字体，使文本的展示具有可读性、易读性、诱目性的艺术和技术。这种安排包含选择字体，设定字号、行长、行间距，以及字母间距。术语西文排印也会用于在此过程中产生的字母、数字、符号的风格、安排、外观。字体设计与西文排印关系密切，有时也会作为西文排印的一部分。大部分西文排印师并不设计字体，而一些字体设计师并不认为自己是西文排印师。西文排印也会被用来指与信息传达无关的装饰图案。|

| WHAT IS
TYPOGRA-
PHY | 任何涉及到安排单词、字母、数字和符号以出版、展示的工作，都在西文排印的范围内，比如组版师、排印师、平面设计师、艺术从业者、动漫师、涂鸦师，甚至公司文员、报刊作者等。电子时代之前，西文排印一直是一个专业领域。数字化把西文排印扩大到与设计无关的新人群。随着西文排印的能力变得越来越普遍，由专业西文排印者代代相传的规则应用与训练，变得越来越弱。|

| 西文排印是一种通过安排字体，使文本的展示具有可读性、易读性、诱目性的艺术和技术。 | 字体设计与西文排印关系密切，有时也会作为西文排印的一部分。大部分西文排印师并不设计字体，而一些字体设计师并不认为自己是西文排印师。西文排印也会被用来指与信息传达无关的装饰图案。西文排印是一种通过安排字体，使文本的展示具有可读性、易读性、诱目性的艺术和技术。这种安排包含选择字体，设定字号、行长、行间距，以及字母间距。术语西文排印也会用于在此过程中产生的字母、数字、符号的风格、安排、外观。

数字化把西文排印扩大到与设计无关的新人群。随着西文排印的能力变得越来越普遍，由专业西文排印者代代相传的规则应用与训练， |

15

5. 形式化实验。1 倍行距也适用汉字方格，但会让汉字彻底变为文字形成的块面，缺点是大量文字会导致阅读困难，适合少量文字的排印。如果字体用细体，西文用大写，加入形式化的、无直接意义的几何装饰图形，涂上基本色，会发现变为了当下很流行的排版风格。

什么是
西文排印

WHAT IS TYPOGRA-PHY

西文排印是一种通过安排字体，使文本的展示具有可读性、易读性、诱目性的艺术和技术。这种安排包含选择字体、设定字号、行长、行间距，以及字母间距。术语西文排印也会用于在此过程中产生的字母、数字、符号的风格、安排、外观。字体设计与西文排印关系密切，有时也会作为西文排印的一部分。大部分西文排印师并不设计字体，而一些字体设计师并不认为自己是西文排印师。西文排印也会被用来指与信息传达无关的装饰图案。

任何涉及到安排单词、字母、数字和符号以出版、展示的工作，都在西文排印的范围内，比如组版师、排印师、平面设计师、艺术从业者、动漫师、涂鸦师，甚至公司文员、报刊作者等。电子时代之前，西文排印一直是一个专业领域。数字化把西文排印扩大到与设计无关的新人群。随着西文排印的能力变得越来越普遍，由专业西文排印者代代相传的规则应用与训练，变得越来越弱。

西文排印是一种通过安排字体，使文本的展示具有可读性、易读性、诱目性的艺术和技术。

字体设计与西文排印关系密切，有时也会作为西文排印的一部分。大部分西文排印师并不设计字体，而一些字体设计师并不认为自己是西文排印师。西文排印也会被用来指与信息传达无关的装饰图案。西文排印是一种通过安排字体，使文本的展示具有可读性、易读性、诱目性的艺术和技术。这种安排包含选择字体、设定字号、行长、行间距，以及字母间距。术语西文排印也会用于在此过程中产生的字母、数字、符号的风格、安排、外观。

数字化把西文排印扩大到与设计无关的新人群。随着西文排印的能力变得越来越普遍，由专业西文排印者代代相传的规则应用与训练，

15

隐藏网格后的排版形式。

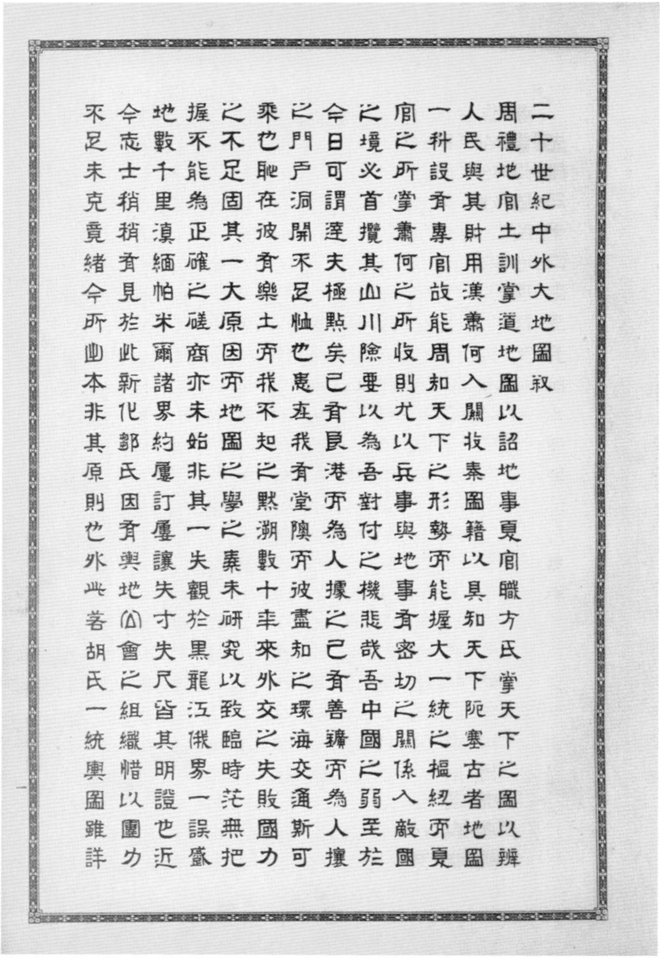

> 二十世紀中外大地圖叙
>
> 周禮地官土訓掌道地圖以詔地事邍司職方氏掌天下之圖以辨
> 人民與其財用漢蕭何入關收秦圖籍以具知天下阨塞古者地圖
> 一科設有專官故能周知天下之形勢而能握大一統之樞紐而畏
> 肩之所掌肅何之所收則尤以兵事與地事為密切之關係入敵國
> 之境必首攬其山川險要以為吾對付之機悲哉已有員中國之弱至於
> 乘之樂堂陛之間為人擄而為人擄斯為人擄
> 之可謂達夫極點矣吾已有善諸而已為人擄
> 今日户洞開在波竟不足恤也近我有堂陛之间波盖十年來外交之失敗國力
> 之不固其一大原因亦未始非其一之黑幕江俄界約壘讓失寸失尺皆其明證也
> 握不能為正確之磋商亦未約壘訂壘讓失寸失尺皆其明證也
> 地戰千里滇緬帕米爾諸界約壘訂壘讓失寸失尺皆其明證也
> 今志士稍稍有見於此新化鄒氏因有興地公會之組織惜以團力
> 不足未克竟緒今所刊本非其原則也外此著胡氏一統輿圖雖詳

继续试验。在汉字排印上，古代有一种稀疏的点阵式形式，汉字与汉字之间间隔很远。Fig.112 是笔者收藏的一本光绪三十二年《二十世纪中外大地图》里的序言页，字体用手写的隶书，每个汉字周围形成方形的均匀空白，使复杂的汉字产生了独特的美感。

113

王序

吾國人士之治外國語、以習英語者爲最多、蓋以貿易額言、英美合計、常居首位、而外邦先進之滿載學術以灌輸於我國者、亦以英美人士爲最勤、顧英語教育之在我國、自最初輸入以迄於今、得分爲三時期、自科舉制度未廢前爲第一期、其時之英語教育、以造就譯人爲目的、或爲外交官吏之預備、或爲貿易往來之媒介、蓽路藍縷、其功匪易、其術未精、此一時也、自學校制度已頒後爲第二期、其時之英語教育、以預備直接閱書爲目的、故大部分之英語教授、胥於中學校行之、備其畢業以後、負笈游學、或轉入專門學校、得以英文原籍、直接講習、而無阻滯難通之病、此又一時也、自最近四五年間爲第三期、蓋就貿易關係言、從前祇需略諳酬對之語言、而今後漸有直接貿易之趨勢、就國交關係言、從前僅需外交官書之鞮寄、而今後則有待國民親交之輔助、自學術界趨向言、從前唯求汲取彼邦學術以藥本國之貧陋、而今後則有賴於溝通彼我思想、以融成適應時勢之文化、英語教育之主旨、既與時推移、則英語教育所恃以推行之字書、其編訂方針、亦自當有異於昔、予懷此感想、已歷多時、今商務印書館馳書抵予、以所纂英漢雙解韋氏大學字典見示、謂不日出版、囑弁一言、予受而讀之、輒深幸出版界中、果有起而彌此缺憾者也、韋書價值、在彼邦學界、久有定評、而編纂精審、尤適中土人士之用、以予視之、厥有數端、各科學術、類有專名、韋書搜羅百籍、所下解詁、悉有來源、剖示詳明、精而不晦、專攻之士、既便檢尋、通常學子、亦竊崖畧、引人入勝、濟此學荒、一也、英語難字、不在玄名、而每在尋常習見之字、蓋義解相生、寖寖益多、愈轉愈遙、不易辨憶、韋書依次列解、示其引申、明其轉注、尋條索脈、觸類旁通、款竅既得、修習匪難、二也、英語字彙、每有同義數字、例得通假、而內含稍別、僅差釐黍、韋書特列比較、剖析毫芒、各字義蘊、宜然畢達、學者翻檢、易使思想趨於邃密、三也、商務印書館取此巨著、加以漢譯、選擇之當、固無間然、而予所尤樂爲贊美者、從前編訂字書、往往去取增刪、不依一本、是編悉照原書、依次譯註、唯達唯雅、尤重在信、閱者對照英漢、可證彼我文字之異同、而得相當轉譯之原字、於隸通思想、最有裨益、此蓋英語教育演進至第三期後最適用之字書、不僅學海鑿幽之導炬、亦增進國民親交之利器矣、並世覽者、或不以予言爲妄歟、民國十一年十一月

王寵惠

(x)

在铅字印刷时代，也有类似的形式。Fig.113 是作者收藏的一本 1923 年的《英汉双解韦氏大学字典》里的序言页，汉字与汉字各成一点，铅字更规整。每个汉字都像星星一样，构成一片天空，美不胜收。

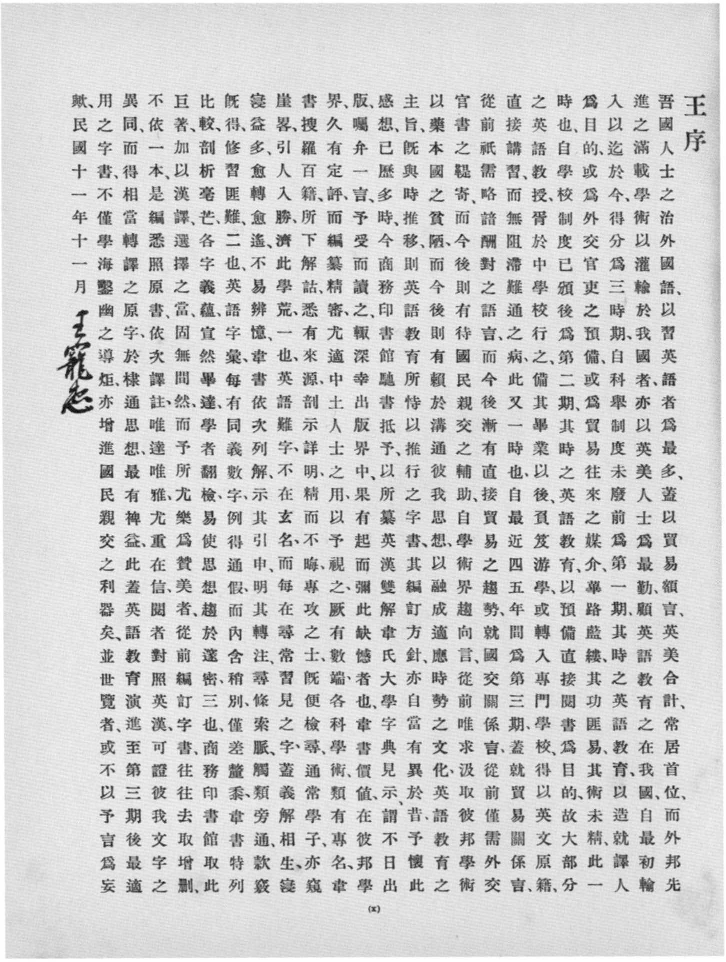

什么是西文排印	西	文	排	印	是	一	种	通	过	安
	排	字	体	，	使	文	本	的	展	示
	具	有	可	读	性	、	易	读	性	、
	诱	目	性	的	艺	术	和	技	术	。
	这	种	安	排	包	含	选	择	字	
	体	，	设	定	字	号	、	行	长	、
What is Typography	行	间	距	，	以	及	字	母	间	
	距	。	术	语	西	文	排	印	也	会
	用	于	在	此	过	程	中	产	生	的
	字	母	、	数	字	、	符	号	的	风
	格	、	安	排	、	外	观	。	字	体
	设	计	与	西	文	排	印	关	系	密
	切	，	有	时	也	会	作	为	西	文
	排	印	的	一	部	分	。	大	部	分
	西	文	排	印	师	并	不	设	计	字
	体	，	而	一	些	字	体	设	计	师
	并	不	认	为	自	己	是	西	文	排
15	印	师	。	西	文	排	印	也	会	被

先用最简单的模型，字号为 8pt，行距为 16pt，字间距设定为 1 000，这时字与字之间的间隔为一个汉字，仍然适用汉字方格。

什么是西文排印	西文排印是一种通过安排字体，使文本的展示具有可读性、易读性、诱目性的艺术和技术。这种安排包含选择字体，设定字号、行长、行间距，以及字母间距。术语西文排印也会
What is Typography	用于在此过程中产生的字母、数字、符号的风格、安排、外观。字体设计与西文排印关系密切，有时也会作为西文排印的一部分。大部分西文排印师并不设计字体，而一些字体设计师并不认为自己是西文排
15	印师。西文排印也会被

隐藏网格后的形式。一般少量文字可以疏排，大量文字可以密排，这里反过来了。左边的密排标题，呈线状；右边的稀疏正文，呈点阵化排列，整体又形成面，层次很丰富。同时弱化了大量汉字排版的不透气感。当然阅读体验会差一点，适用于偏重形式感，或者文字数量中等的版式。

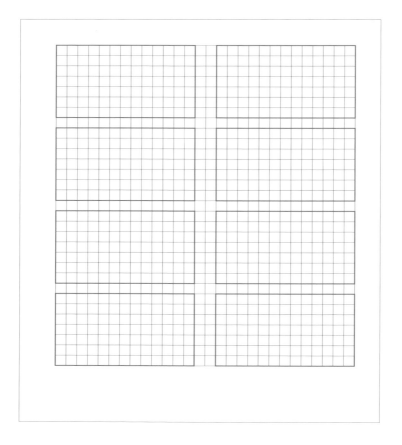

2×4 的网格。除了 3×4 的网格，还有很多其他可能性。

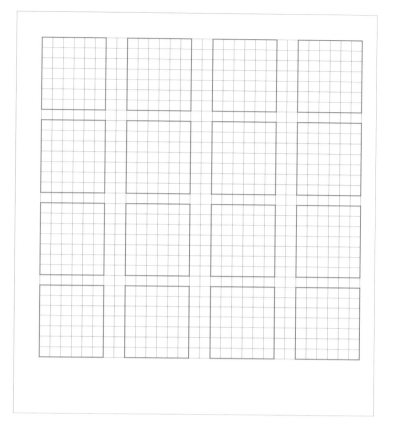

4×4 的网格，文字的设定类似之前的方法。

什么是西文排印
What is Typography

西文排印是一种通过安排字体，使文本的展示具有可读性、易读性、诱目性的艺术和技术。这种安排包含选择字体，设定字号、行长、行间距，以及字母间距。术语西文排印也会用于在此过程中产生的字母、数字、符号的风格、安排、外观。字体设计与西文排印关系密切，有时也会作为西文排印的一部分。大部分西文排印师并不设计字体，而一些字体设计师并不认为自己是西文排印师。

任何涉及到安排单词、字母、数字和符号以出版、展示的工作，都在西文排印的范围内，比如组版师、排印

西文排印是一种通过安排字体，使文本的展示具有可读性、易读性、诱目性的艺术和技术。

15

形式化实验，增加变化。文字由 1 栏变为 2 栏，并错落分布；正文并不排满，增大留白；标题字号增大至 20pt，增强与正文的对比。

什么是西文排印
What is Typography

西文排印是一种通过安排字体，使文本的展示具有可读性、易读性、诱目性的艺术和技术。

15

西文排印是一种通过安排字体，使文本的展示具有可读性、易读性、诱目性的艺术和技术。这种安排包含选择字体，设定字号、行长、行间距，以及字母间距。术语西文排印也会用于在此过程中产生的字母、数字、符号的风格、安排、外观。字体设计与西文排印关系密切，有时也会作为西文排印的一部分。大部分西文排印师并不设计字体，而一些字体设计师并不认为自己是西文排印师。

任何涉及到安排单词、字母、数字和符号以出版、展示的工作，都在西文排印的范围内，比如组版师、排印

Chinese grid system

隐藏网格后的排版形式。

排版是一种通过安排字体，使文本的展示具有可读性、易读性、诱目性的艺术和技术。这种安排包含选择字体，设定字号、行长、行距，以及字母间距。字体设计与排版关系密切。

排版是一种通过安排字体，使文本的展示具有可读性、易读性、诱目性的艺术和技术。

排版是一种通过安排字体，使文本的展示具有可读性、

114

01. 什么是排版
02. 什么是排版
03. 什么是排版
04. 什么是排版
05. 什么是排版
06. 什么是排版
07. 什么是排版
08. 什么是排版
09. 什么是排版
10. 什么是排版
11. 什么是排版
12. 什么是排版
13. 什么是排版
14. 什么是排版
15. 什么是排版
16. 什么是排版
17. 什么是排版
18. 什么是排版

01. 什么是排版
02. 什么是排版
03. 什么是排版
04. 什么是排版
05. 什么是排版
06. 什么是排版
07. 什么是排版
08. 什么是排版
09. 什么是排版
10. 什么是排版
11. 什么是排版
12. 什么是排版

01. 什么是排版
02. 什么是排版
03. 什么是排版
04. 什么是排版
05. 什么是排版
06. 什么是排版
07. 什么是排版
08. 什么是排版
09. 什么是排版

115

1.8 倍行距的中文网格

1. 开本近似 A4，210mm×280mm，字号基数设定为 3pt，行距设定为 1.8 倍。正文字号为 3×3pt=9pt，行距为 1.8×9pt=16.2pt；标题字号为 4×3pt=12pt，行距为 1.8×12pt=21.6pt；注解字号为 2×3pt=6pt，行距为 1.8×6pt=10.8pt(Fig.114)。这样标题、正文、注解的字号比为 4:3:2，同样行距比也为 4:3:2。表现在排版上，2 行正文与 3 行注解高度相同，而它们的字号比为 3:2；4 行正文与 3 行标题高度相同，而它们的字号比为 3:4；2 行注解与 1 行标题高度相同，而它们的字号比为 2:1（Fig.115）。整个文本形成系统的数学韵律。

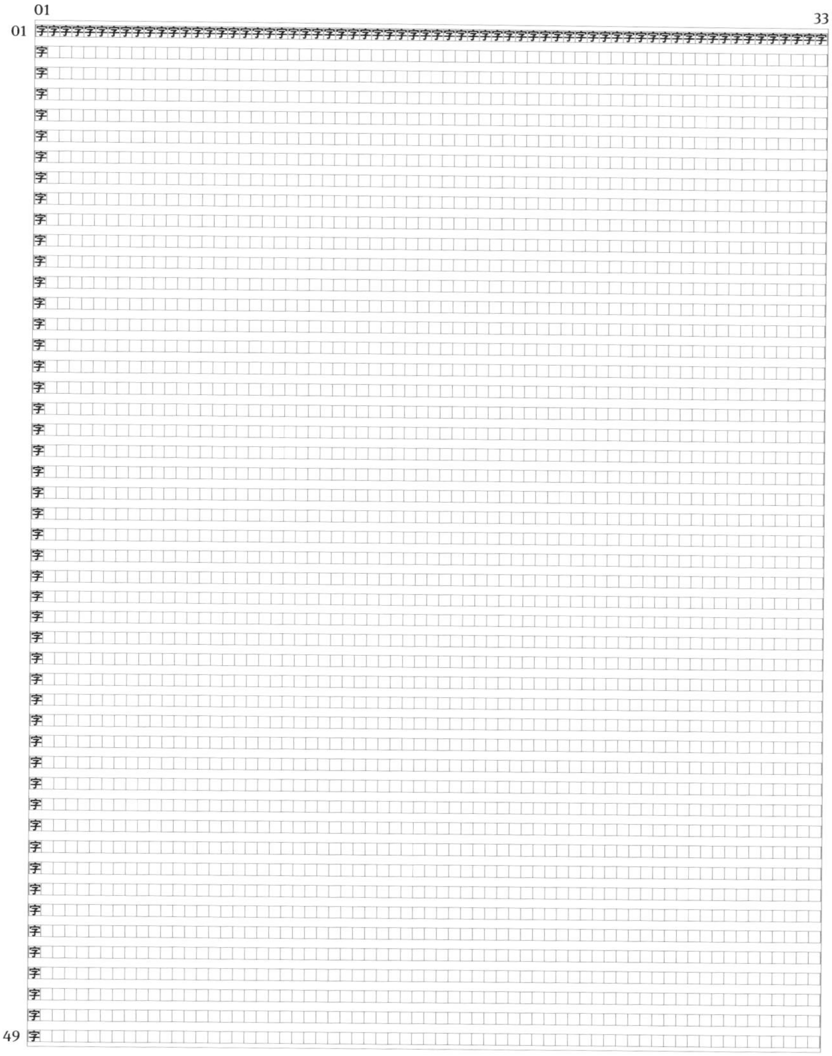

2. 用正文字号 9pt、行距 16.2pt 的单元格填满页面。上图为原页面的 50%，最多容纳 49 行，每行最多容纳 33 个字。

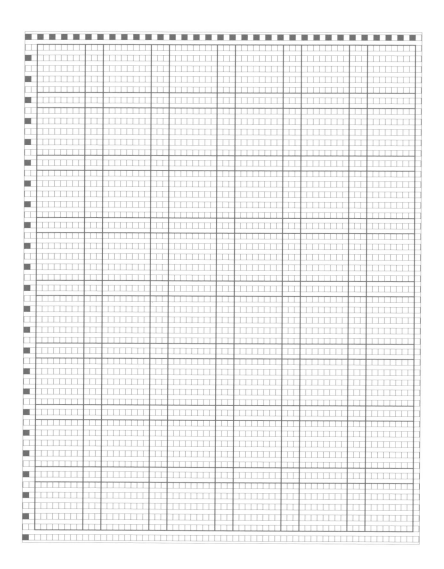

3. 根据排版需要设置网格，方法类似于国际风格的网格设定，这里以 6×8 的网格为例。上图是字容量最大的 6×8 网格，每个单元格可容纳 5 行，每行容纳 8 个字。版心几乎撑满页面，是偏流行风格的版心设置。之后以此为基础，通过减少字容量，可得到不同风格的 6×8 网格。

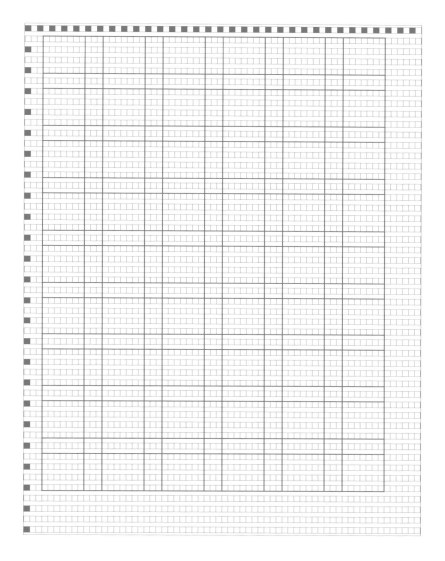

4. 同样是 6×8 的网格,单元格缩小一些,每个单元格可容纳 4 行,每行容纳 7 个字。留白基于功能需要,是偏国际风格的版心设置。用此网格设计封面。

字号从小到大依次为：3×3pt=9pt、4×3pt=12pt、6×3pt=18pt、15×3pt=45pt，整个页面都用 1.8 倍行距。这时视觉上西文行距过大，与中文行距的韵律明显不同。可把中文放大到与西文同样大小，然后根据中文的行距调整西文的行距。

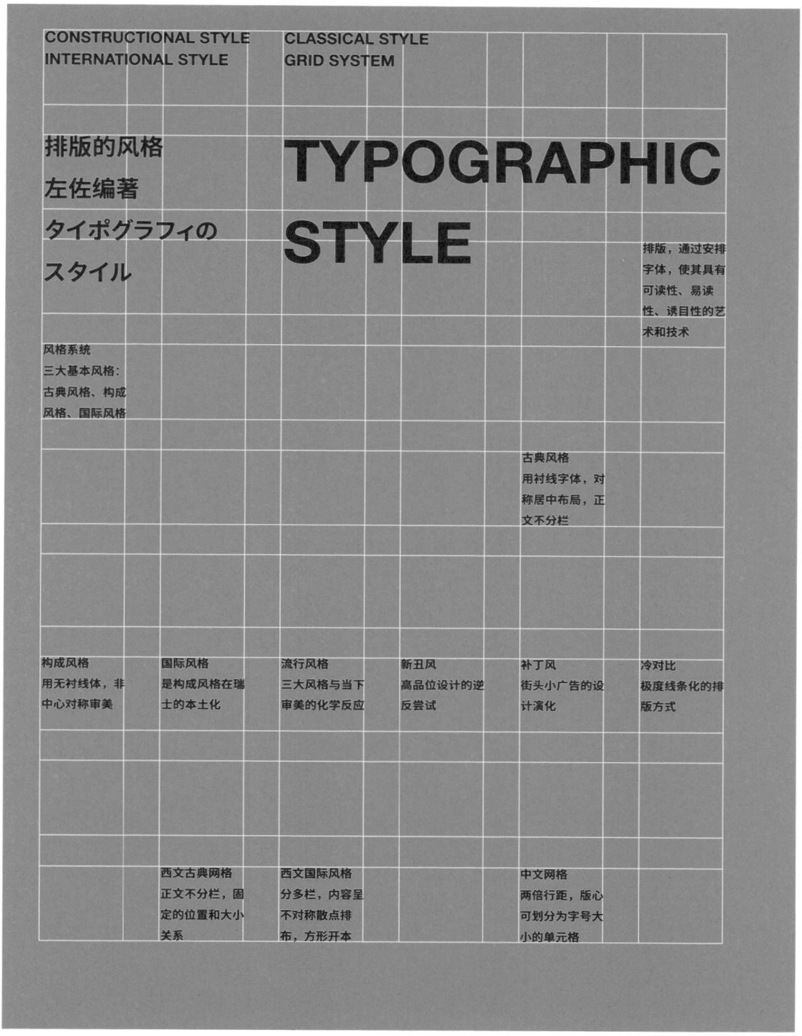

Chinese grid system

| CONSTRUCTIONAL STYLE | CLASSICAL STYLE |
| INTERNATIONAL STYLE | GRID SYSTEM |

TYPOGRAPHIC STYLE

排版的风格
左佐编著
タイポグラフィの
スタイル

排版，通过安排字体，使其具有可读性、易读性、谕目性的艺术和技术

风格系統
三大基本风格：
古典风格、构成风格、国际风格

古典风格
用衬线字体，对称居中布局，正文不分栏

| 构成风格 | 国际风格 | 流行风格 | 新丑风 | 补丁风 | 冷对比 |
| 用无衬线体，非中心对称审美 | 是构成风格在瑞士的本土化 | 三大风格与当下审美的化学反应 | 高品位设计的逆反尝试 | 街头小广告的设计演化 | 极度线条化的排版方式 |

| | 西文古典网格 | 西文国际风格 | | 中文网格 | |
| | 正文不分栏，固定的位置和大小关系 | 分多栏，内容呈不对称散点排布，方形开本 | | 两倍行距，版心可划分为字号大小的单元格 | |

调整西文行距后的排版形式。版式也使用了构成风格的方形法则。比如，中间大面积的空白是一个方形，文字块与文字块之间也形成了很多隐藏的方形。

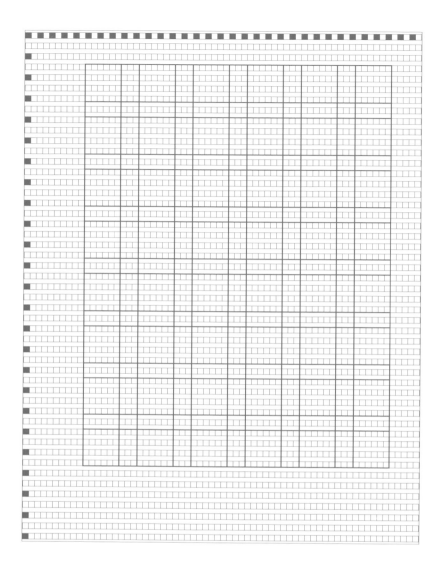

5. 同样是 6×8 的网格，单元格继续缩小，每个单元格可容纳 4 行，每行容纳 6 个字。页边距留白很大，是偏古典风格的版心设置。书籍《治字百方》内页即是用的此网格。

本书的网格

开本设定为 420pt×670pt，长宽比接近黄金比例，横向划分为 42 个 10pt 的单元格，纵向划分为 67 个 10pt 的单元格。左侧页边距为 50pt，右侧页边距为 30pt，天头为 30pt，地脚为 90pt，版心为 340pt×550pt，页码距底部 50pt。左右页的左右页边距不做对称处理。

版心划分为 5 个间隔 10pt 的纵列，纵向上以 10pt 的单元格自由决定。正文字号为 10pt，行间距为 20pt，这样正好匹配整个 10pt 的单元格系统。段首不缩进，段落之间以空行来划分。

本书西文使用 Meta Serif 和 Future Bold 字体，中文使用 Noto Serif 思源宋体与 Noto Sans 思源黑体。

Chinese grid system

内容简介

本书从宏观的角度总结出排版的三大基本风格，古典风格、构成风格与国际风格，建立了一个风格系统，并详细介绍了各自的网格系统。

未经许可，不得以任何方式复制或抄袭本书的部分或全部内容。
版权所有，侵权必究。

图书在版编目（CIP）数据

排版的风格 / 左佐编著 . -- 北京：电子工业出版社，2019.4
ISBN 978-7-121-36119-7
I.①排… II.①左… III.①排版 - 研究 IV.①TS812
中国版本图书馆 CIP 数据核字（2019）第 043887 号

责任编辑：赵英华
印　　刷：北京尚唐印刷包装有限公司
装　　订：北京尚唐印刷包装有限公司
出版发行：电子工业出版社
　　　　　北京市海淀区万寿路 173 信箱　邮编 100036
开　　本：720×1000　1/16　印张：10.5　字数：272.8 千字
版　　次：2019 年 4 月第 1 版
印　　次：2023 年 4 月第 5 次印刷
定　　价：89.00 元

凡所购买电子工业出版社图书有缺损问题，请向购买书店调换。若书店售缺，请与本社发行部联系，联系及邮购电话：(010) 88254888，88258888。

质量投诉请发邮件：zlts@phei.com.cn，盗版侵权举报请发邮件：dbqq@phei.com.cn。

本书咨询联系方式：(010) 88254161~88254167 转 1897